Microscopy: A Very Short Introduction

VERY SHORT INTRODUCTIONS are for anyone wanting a stimulating and accessible way into a new subject. They are written by experts, and have been translated into more than 40 different languages.

The series began in 1995, and now covers a wide variety of topics in every discipline. The VSI library now contains over 350 volumes—a Very Short Introduction to everything from Psychology and Philosophy of Science to American History and Relativity—and continues to grow in every subject area.

Very Short Introductions available now:

SLEEP Steven W. Lockley and
      Russell G. Foster
SOCIAL AND CULTURAL
      ANTHROPOLOGY
      John Monaghan and Peter Just
SOCIALISM Michael Newman
SOCIOLINGUISTICS John Edwards
SOCIOLOGY Steve Bruce
SOCRATES C. C. W. Taylor
THE SOVIET UNION Stephen Lovell
THE SPANISH CIVIL WAR
      Helen Graham
SPANISH LITERATURE Jo Labanyi
SPINOZA Roger Scruton
SPIRITUALITY Philip Sheldrake
SPORT Mike Cronin
STARS Andrew King
STATISTICS David J. Hand
STEM CELLS Jonathan Slack
STRUCTURAL ENGINEERING
      David Blockley
STUART BRITAIN John Morrill
SUPERCONDUCTIVITY
      Stephen Blundell
SYMMETRY Ian Stewart
TAXATION Stephen Smith
TEETH Peter S. Ungar
TERRORISM Charles Townshend
THEATRE Marvin Carlson

THEOLOGY David F. Ford
THOMAS AQUINAS Fergus Kerr
THOUGHT Tim Bayne
TIBETAN BUDDHISM
      Matthew T. Kapstein
TOCQUEVILLE Harvey C. Mansfield
TRAGEDY Adrian Poole
THE TROJAN WAR Eric H. Cline
TRUST Katherine Hawley
THE TUDORS John Guy
TWENTIETH-CENTURY
      BRITAIN Kenneth O. Morgan
THE UNITED NATIONS
      Jussi M. Hanhimäki
THE U.S. CONGRESS Donald A. Ritchie
THE U.S. SUPREME COURT
      Linda Greenhouse
UTOPIANISM Lyman Tower Sargent
THE VIKINGS Julian Richards
VIRUSES Dorothy H. Crawford
WITCHCRAFT Malcolm Gaskill
WITTGENSTEIN A. C. Grayling
WORK Stephen Fineman
WORLD MUSIC Philip Bohlman
THE WORLD TRADE ORGANIZATION
      Amrita Narlikar
WORLD WAR II Gerhard L. Weinberg
WRITING AND SCRIPT
      Andrew Robinson

Available soon:

LIBERALISM Michael Freeden
INFECTIOUS DISEASE
      Marta L. Wayne and
      Benjamin M. Bolker

SOCIAL WORK Sally Holland and
      Jonathan Scourfield
FORESTS Jaboury Ghazoul
PSYCHOANALYSIS Daniel Pick

For more information visit our website

www.oup.com/vsi/

Terence Allen

MICROSCOPY

A Very Short Introduction

OXFORD
UNIVERSITY PRESS

Great Clarendon Street, Oxford, OX2 6DP,
United Kingdom

Oxford University Press is a department of the University of Oxford.
It furthers the University's objective of excellence in research, scholarship,
and education by publishing worldwide. Oxford is a registered trade mark of
Oxford University Press in the UK and in certain other countries

Published in the United States of America by Oxford University Press
198 Madison Avenue, New York, NY 10016, United States of America

British Library Cataloguing in Publication Data

Data available

Library of Congress Control Number: 2014956893

ISBN 978-0-19-870126-2

Printed and bound by
CPI Group (UK) Ltd, Croydon, CR0 4YY

For Gabrielle and the family

Contents

List of illustrations

Microscopy

Chapter 1
Microscopy and the discovery of a new world

Microscopes and telescopes are means of extending what we can see with the naked eye. Telescopes bring distant items closer, and microscopes make small objects larger. The laws of physics are exactly the same for both, so what are the differences between them? Telescope optics are arranged to collect as much light as possible from afar, whereas microscope optics work from millimetres away from the specimen. For telescopes the objects of interest are by nature completely inaccessible (apart from the odd space probe), whereas specimens for microscopy can be prepared to their best advantage, allowing many options for optimizing structural information. Microscopes and telescopes both appeared around the beginning of the 17th century, as developments from the glass lenses that were used in spectacles. In 1609, perhaps as one of his lesser known achievements, Galileo Galilei demonstrated his *occhiolino* ('little eye') to the Accademia dei Lincei where Giovanni Faber is credited with suggesting it should be called a 'microscope'. This was appropriate as the same group had previously coined the name for a telescope. Thus from similar beginnings, half a millennium ago, telescopes and microscopes continue to explore the structure of the world about us, from the outer reaches of the universe to the visualization of atomic structure.

A good magnifying glass produces an image with a magnification of four or five times the size of the object of interest. Modern light microscopes can increase this magnification to two thousand. Move to electron microscopy and magnification can be increased to around one million, enough to allow direct visualization of individual atoms. Magnification alone, however, is no guarantee of increased information, and any increase in magnification should incorporate enhanced resolution—the ability to see detail; otherwise we see a larger version but no new information—termed 'empty magnification'. Regardless of magnification and resolution, information from the microscope is also dependent on the illumination, be it light or electrons, and their interactions with the specimen to produce contrast in the image. Without the stains discovered by Robert Koch and colleagues mid 19th century, we could still be ignorant of the bacteria which cause diseases such as anthrax and tuberculosis. Alternatively, contrast can be generated by manipulation of the optical pathway to allow direct observation of living cells. Add a recording system, and the wonders of cell division can be followed by time lapse. For the next advance, the pathways of individual molecules within living cells can be tracked using fluorescence microscopy. These particular milestones are taken from life sciences, but similar progress has been made across the whole range of microscopic investigation. The development of microscopy is one in which new discoveries in apparently unrelated fields are seized upon and rapidly integrated, for instance lasers, computers, and digital imaging. Entirely new inventions also appear with a regular frequency, such as atomic force microscopy (AFM), which forms images by touch rather than vision, and has become the cornerstone of the field of nanotechnology. Although we may think of a microscope as no more than a relatively simple instrument that we looked down to see a cheek scraping of our own cells at school, microscopy impacts on virtually all aspects of our daily lives. This *Very Short Introduction* will attempt to survey the diverse field of science which is modern microscopy.

Our best ability to see detail with the eye is to perceive two strands of human hair that are separated by a hair's width. This is the limit of our resolution—the ability to see detail, measured by the distance at which two points are still distinct. Magnification without increased resolution merely produces a larger image with no increase in detail. Amazing as our own eyes are, they are poor compared with those of an eagle, where the resolution is eight times as good, enabling it to spot a rabbit at a distance of two miles.

In order to make things bigger, a magnified image must be produced. This requires a transparent material which 'bends' or refracts light. Water refracts light—perhaps not the first thing that comes to mind—but at the beginnings of television, water filled 'magnifiers' were a routine domestic requirement for the tiny screens of the day. However, from the very beginnings of microscopy, glass has been the standard material.

Lenses and the first instruments

Glass is thought to have been created first in the bronze age (around 3000 BC), and both glass beads and glass mosaics were in use in Egypt from around 2500 BC. As an aid to vision, 'reading stones' (glass spheres laid directly on top of reading materials) first appeared around AD 1000. The first wearable eyeglasses were produced in Italy, in around AD 1300. Some three hundred years later, a Dutch father and son, Zaccharias and Hans Jensen, placed lenses at the ends of an adjustable tube which produced a magnified image, thus inventing the forerunner of both microscopes and telescopes. At the start of the 17th century, Galileo Galilei, then a professor of mathematics at the University of Padua, produced a compound (two-lens) microscope which he called his *occhiolino* (little eye), and presented it to Prince Frederico Cesi, founder of the Accademia dei Lincei—a learned society that pre-dated the UK Royal Society in Britain by half a century. The Linceans (self-styled 'lynx-eyed') were responsible for creating the names of both telescope and microscope, the latter by Giovanni Faber, from the Greek *micron* (meaning 'small') and *skopein* ('to look at').

In England, the newly formed Royal Society (1660) appointed Robert Hooke as its first Curator of Experiments. Hooke, the Gresham Professor of Geometry, was also a prominent architect and a surveyor involved in the reconstruction after the Great Fire of London. He too built telescopes as well as a compound microscope (Figure 1a). In 1665, in *Micrographia* (the second book published by the Royal Society), Hooke likened the structure of plant tissues to the cells inhabited by monks, thus naming the basic unit of life. Again, as with Galileo, Hooke was a polymath, making notable contributions to many other branches of science.

Antoni van Leeuwenhoek

Although compound microscopes had existed for half a century, the magnification they produced was modest, and the best magnification available was about thirty times. The full impact of visualizing the living world beyond this level was brought about by a linen draper from Delft called Antoni van Leeuwenhoek. As an apprentice in a dry goods store, Leeuwenhoek would have used magnifying lenses to count the threads in cloth. Subsequently he became expert in producing single spherical lenses from drops of molten glass which gave magnifications of up to 266 times, and his microscopes (Figure 1b) along with Hooke (Figure 1a) were to reveal an astounding world of organisms never even imagined. Leeuwenhoek produced hundreds of microscopes in his life which varied considerably in performance and attainable magnification, suiting some specimens better than others. Despite very little formal education, and writing his reports in Low Dutch, Leeuwenhoek sent most of his discoveries to the Royal Society of London, where they were received with an admirable open mindedness and recognition for his genius in microscopy. He sent over 200 papers over a period of nearly half a century, and in 1680 was elected a member. Leeuwenhoek obviously made clearer lenses than others, but also worked extremely hard to produce his drawings—to quote: 'on the close inspection of 3 or 4 drops, I may indeed expend so much labour that sweat breaks out on me.' Certainly

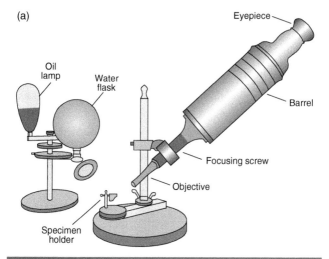

1. (a) Hooke's microscope is a compound microscope with two lenses, whereas (b) Leeuwenhoek's instrument had a single spherical lens.

with one eye squinting into the tiny lens with the instrument resting on his cheek or forehead, and constantly moving to get the best light to image his specimens, Leeuwenhoek had things much harder than the modern microscopist, who sits in a comfortable chair and looks down two eyepieces through sophisticated lenses at an optimally illuminated specimen on a rigid and mechanically driven stage.

Leeuwenhoek's results caught the imagination as no microscopy had done before. The idea that there was a whole world of life in a drop of pond water was met with total astonishment. He observed bacteria and yeast for the first time, red blood cells and human sperm, diatoms, and virtually anything which he could attach to his microscope and view, often using tiny tubes of glass to contain specimens in water. Leeuwenhoek has been called the father of microbiology, but he also laid the foundations of plant anatomy and animal reproduction, suggesting that fertilization occurred when the sperm penetrated the egg. His total production of microscopes was around five hundred, of which nine are still known to exist.

A scientific approach—Carl Zeiss

From the end of the 17th century until the mid 19th century, the world of microscopy seems to have marked time. A significant change occurred when Carl Zeiss, a 30-year-old mechanic, set up a scientific instrument workshop in Jena, Germany. Compound microscopes were in demand in science, but the lenses were still manufactured on a trial and error basis. In 1866, in an attempt to incorporate basic science into a standardized production, Zeiss approached Ernst Abbe, a physicist from the local university. After five years of work, Abbe established that increasing the size of the lens aperture improves the function of the microscope, and that the wavelength of light limits the recognition (resolution) to those structures that are larger than half the wavelength. Abbe worked out the mathematical formula for the conditions to be fulfilled for sharp imaging, which, together with improvements in glass technology from the glass chemist Otto Schott, made it possible to produce microscopes of the highest optical quality. Although Abbe had 'only' demonstrated a basic truth in physics that the wavelength of light was the limiting factor in resolution, such was the acceptance of this 'diffraction barrier' that it was to be over a hundred years before ways to get round it started to emerge.

Abbe became an equal partner with Zeiss and business boomed, and by the end of the 1880s, the Zeiss workforce totalled some 360 people. When Zeiss died in 1888, Abbe established the Zeiss Foundation, funding research and new products, and also made his mark as a social reformer, creating conditions of service for his workers with benefits way ahead of their time. The last significant piece of this optical jigsaw was provided by August Kohler, a member of Zeiss's staff who worked out the best way to illuminate the specimen, implementing the total resolving power of Abbe's objective lenses. Setting up Kohler illumination (making sure the light path is centred and focused) is still the first thing any serious microscopist does when sitting at a microscope. Despite two world wars, the partition and reunification of Germany, and the rise of Far Eastern camera and microscope technology and manufacture, Zeiss has survived to this day as a significant player in microscopy. Zeiss was not, however, the only German manufacturer of high-quality microscopes, with both Leitz and Wild and also Reichert (in Austria), all of whom were renowned producers of research grade instrumentation. All three are now under the brand name of Leica.

Microscopy in the 20th century

From the beginning of the 20th century, further improvements and major advances in the field of microscopy have come at an increasingly rapid pace compared to its first three centuries. Much of the 'raw meat' of these advances has come from the incorporation of new aspects of unrelated technology, a feature of microscopy which continues to characterize its development up to the present day. Examples include novel light sources, such as lasers and light emitting diodes, and the latest in image recording technology (from film to digital acquisition).

In 1903, Richard Zsigmondy reported his studies of gold colloids, suspensions of tiny particles which produced different colours according to the particle size, and although he did not know it at the time, he was repeating earlier studies by Michael Faraday, reported to the Royal Society in 1847. In order to visualize particles well below Abbe's half wavelength barrier, he invented an ingenious system of illumination that gave a dark background against which visible cones of light emanated from individual gold particles, some as small as 10 nanometres. The light spots Zsigmondy saw against a black background (called Tyndal cones) in his otherwise relatively conventional microscope were much bigger than the particles themselves, but his ultramicroscope still 'broke' the diffraction barrier in visualizing particles less than half the wavelength of light. For his work on colloidal particles, Zsigmondy received the Nobel Prize for Chemistry in 1925. As colloidal gold particles fall in the range of today's 'nanoparticles', he has also been termed (retrospectively) 'The Father of Nanotechnology'. As we shall see (Chapter 6), nearly a century and a half after Faraday, gold nanoparticles reappeared to become standard imaging tags as probes for electron microscopy.

Phase and fluorescence microscopy

Frits Zernike's invention of phase contrast microscopy revolutionized the microscopy of living biological systems. Living cells are mostly colourless, and therefore have little or no inherent contrast. Different parts of cells will react differentially with many stains, as had been shown by early microscopical anatomists such as Golgi and Cajal. Staining usually requires the cell to be chemically preserved (fixed), although there are a small number of 'vital stains' which will work with living cells. In 1930, Zernike, a Dutch physicist, was working on diffraction gratings, and realized that the small differences in the refractive index of different parts of cells produced a phase shift as the light passed through them. The speed of the light waves was altered, and although our eyes cannot pick up the differences in wave interference, Zernike's invention (using phase retarding rings in both the microscope condenser and objective lens) was able to convert the phase changes into changes in contrast so that some areas within cells appeared darker than others (Figure 2a). This opened the door to

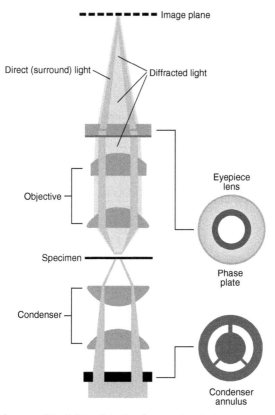

2a. Diagram of the light path in the phase contrast microscope.

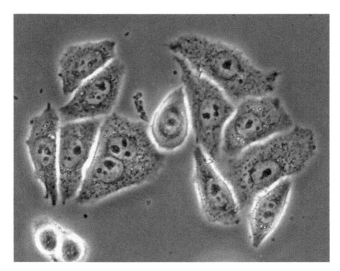

2b. Phase contrast image of living cultured human cells.

live cell imaging (Figure 2b) and Zernike was awarded the Nobel Prize in Physics in 1953. There is, however, an ironic twist to this tale, in that Zernike's discovery went largely unappreciated in the Zeiss factory until 1941, when the German Wehrmacht took control of all potential contributions to the war effort, and decided to produce phase contrast microscopes—at the same time that Zernike and his fellow countrymen were suffering under the German occupation of the Netherlands.

The first fluorescence microscope was built by August Köhler at the Zeiss factory in Jena in 1904, as an attempt to improve the resolution for light microscopy by the use of UV light (which has half the wavelength of visible light). Köhler noted that some specimens generated fluorescence naturally as a result of the UV illumination, but it was Oskar Heimstadt, seven years later, who managed to filter out all but the UV light from the arc lamp source, and used a dark field background to allow the best visualization of fluorescence generated by the specimen. Heimstadt himself was unsure of the impact of his studies, and it was not until a UV system was devised which illuminated specimens from above (epifluorescence), together with the use of fluorescent chemicals (fluorochromes) as stains, that the possibilities which fluorescence offered were actually realized. Forty years later, Johan Sebastian Ploem, a Dutch scientist working with Leitz, produced further advances with the development of optical devices (dichromatic beamsplitters and mirrors) that could reflect a narrow band of wavelengths while transmitting all the others, allowing illumination with a single wavelength. Today (as we shall see in Chapter 2), the emergence of lasers, along with many novel fluorochromes, together with advanced optical and imaging systems, has made fluorescence microscopy a cornerstone of cell biology.

The impact of microscopy on medicine

The second half of the 19th century was an age of enlightenment with respect to the establishment of the true source of many diseases. After Semmelweis had shown that infections during childbirth were largely prevented by hand washing, further progress was made as the result of microscopic observation and scientific experiment. This was the period in which Louis Pasteur demonstrated that spontaneous generation does not occur, confirming his support for the germ theory of disease. Robert Koch, a country doctor, used his microscope to discover the bacteria that caused tuberculosis and cholera, as well as the anthrax bacterium and its life cycle. Paul Ehrlich, a student of Koch, used newly synthesized dyes from the emerging chemical industry to stain and give contrast to specimens in order to enhance the visualization of cell structure in the progression of disease. Christian Gram developed a stain still used to this day to differentiate bacteria (Gram negative or Gram positive). While none of these pioneers of microbiology did anything to actually improve the performance of microscope optics, per se, this was the period in which microscopy became an integral part of the diagnosis, classification, and treatment of disease.

The birth of electron microscopy

The reduction in the wavelength of electrons compared with light offered a massive improvement in resolution, which drove the development of electron microscopy in the first half of the 20th century. The electron itself had been discovered in 1897 by Sir Joseph John Thomson, and its wavelike nature by Sir George Thomson. Louis de Broglie showed (in his Ph.D. thesis) that electrons, like light, behave as both particles and waves, and also established the wavelength, providing a resolution of one thousandth that of light.

Transmission electron microscopy

To observe a specimen with a beam of electrons is straightforward in theory, requiring a source of electrons, the means to focus them on a specimen, and lenses to produce a magnified image and visualize it in some way, as the eye doesn't 'see' electrons. Generating electrons is simple, requiring nothing more than to heat a tungsten wire until it glows, at which point electrons are liberated. In normal atmospheric conditions, they will be absorbed by the first molecule of air that they encounter, so to make any use of them, we have to work in a vacuum. Although a vacuum enables electrons to move freely, they have no impetus to do so, so we create a voltage difference (an accelerating voltage) between their source (the cathode) and an anode some distance away, thus creating a beam of electrons moving in the same direction. This beam will not travel through glass as light does, but it can be focused using electromagnetic lenses, essentially hollow electric coils which produce magnetic fields that converge or diverge the electrons to the same effect that glass lenses modulate light. Even with accelerating voltages of 100,000 volts, electron beams still have poor penetration, meaning that the specimens have to be extremely thin. Finally, having generated our beam, moved it through the specimen, and produced a focused (and magnified) image, we still need to be able to see the result. This was achieved by mounting a fluorescent screen in the plane of focus of the image, and viewing it by eye through a glass window. Electrons of different intensities stimulate the fluorescent coating of the screen, producing a yellow-greenish image, viewed in subdued or dark conditions. Electrons expose photographic emulsion in the same way as light, so a camera (initially glass plates, subsequently film, nowadays an imaging chip) is placed below the fluorescent screen, which is lifted to record exposures.

The first person to fulfil these conditions for electron microscopy and create a practical working instrument was Ernst Ruska, born in Heidelberg in 1906. Ruska had previously worked on electromagnetic lenses, producing a coil with an iron cap called the 'pole shoe' lens which became a standard in electron microscopy. In 1931, with Dr Max Knoll, he produced the first transmission electron microscope (TEM) (electrons passing through the specimen). This was followed by a second instrument in 1933 which gave 'better definition than a light microscope'. Ruska moved into industry, where there were better resources, and after a period of work on television receivers and transmitters, he produced the first commercial electron microscope, made by Siemens in 1939. A year earlier, in the University of Toronto, Franklin Burton and his students Cecil Hall, James Hillier, and Albert Prebus independently produced a functioning instrument. Hillier moved on to RCA, and produced early commercial instruments in the USA, while Siemens produced instruments in Europe.

Despite this success, Ruska couldn't convince industry to invest in the production of his electron microscope. The problem of overheating specimens had not yet been overcome. Ruska had discovered a partial solution to this in that very thin samples could be resolved by electrons through diffraction only, and not absorption, which meant less heat was absorbed. However, this wasn't enough to convince a sceptical industry to part with their money. 'Who are the potential customers of such a device?' they would ask themselves.

The obvious answer was biologists, but what did Ruska's microscope do to biological specimens? The vacuum tube dehydrated them, the electrons damaged them, and the heat that was generated burnt them. All of these apparently insurmountable problems were to be solved by a variety of ingenious solutions (see Chapter 4).

After the Second World War, Ernst Ruska designed the Siemens 'Elmiskop 1' in 1954, which became a worldwide success, with the vast majority still functioning decades after installation. Over fifty years after the building of the first instrument, then aged 80, Ruska was awarded the Nobel Prize in Physics in 1986 (emphasizing the requirement for longevity as a Nobellist), just two years before he died.

Scanning electron microscopy

The idea of scanning electron microscopy (moving a spot of electrons to and fro across the specimen in a zig-zag or rastered pattern) to generate surface information and get round the necessity for ultrathin specimens has been around from the earliest days of electron microscopy. However, practical difficulties and a limited performance expectation delayed production of the first commercial scanning EM until 1964. The instrument, built by the Cambridge Instrument Company, was the result of the vision of Charles Oatley, from the Engineering Department of the University. After working on the development of radar during the war, he arrived at Cambridge in 1947. Oatley followed earlier work by Manfred von Ardenne in Germany who had produced a working instrument destroyed in an air raid. Oatley, in his own words, 'took up the baton' from von Ardenne, working with a series of students each given a 'simple Ph.D. project' which was to 'Build a scanning Electron Microscope'. Despite being told by many of his peers that he was wasting his time, Oatley persisted, and by the mid 1950s, enough progress had been made to launch a commercial instrument, despite the failure of an American version of a scanning EM made by RCA. A prototype was built for the Pulp and Paper Research

Institute in Canada (SEM3) and subsequently in 1965, when the performance had significantly improved, and the possibilities of scanning EM were looked on with more enthusiasm, a batch of five instruments named 'Stereoscan' were produced by the Cambridge Instrument Company and installed at various institutes in the UK and Europe. These instruments were used mainly in materials science, for microelectronics, paper, and fabric investigations, rather than the imaging of biological specimens. Although Cambridge Instruments led the way in scanning EM they were followed just six months later by Japanese Electrical Optical Industries with their JSM 1. The competition between European companies and the Japanese in electron optics continues to this day, although both British companies, Cambridge Instruments and AEI (who manufactured transmission electron microscopes), were relatively early casualties.

From the 1960s onwards, new discoveries in science and technology were being discovered at an ever accelerating pace, such as lasers, charged coupled devices, computer technologies, digital imaging, light emitting diodes, fluorescent probes, green fluorescent protein, and quantum dots. From the point at which these discoveries became practical realities they were invariably incorporated into various aspects of microscopy. For the rest of this volume we will be concerned with the extraordinary power and variety of modern microscopy.

Chapter 2
A spectrum of microscopies

In order to enable the eye to see detail beyond its normal range, we must provide it with a magnified image, illuminated in some way so we can actually visualize it, and with sufficient contrast to pick out separate parts within that image. The detail we can see is ultimately limited by the resolution of the imaging system we choose to use, but resolution without contrast is irrelevant—would we see a green golf ball on the fairway at 200 yards? Here we shall consider the various parameters contributing to image formation across both light and electron microscopy.

Magnification is a straightforward concept—we hold a magnifying glass over small print, and increase in the size of the virtual image produced by the lens allows us to read more easily. Magnification is a term without particular units, e.g. 'tenfold' or 'times' and sometimes 'diameters' to indicate the level or 'power' of the increase in image size. A stated magnification can be notoriously inaccurate in the printed format, as it obviously depends on the size that an image is reproduced—which could be an entire newspaper page, or a tiny snapshot in a corner somewhere. For scientific publications, a micron bar is incorporated into the image, indicating the scale, which then 'automatically' changes with the size of the reproduction. Another way is to quote 'field width', which indicates the actual size of the picture content regardless of the size of reproduction. Serious scientific publications always insist on some form of scale, but newspaper and magazine editors often prefer to save the space (and would probably add that the average reader had little concept of a micrometre anyway). Consequently we may possibly see successive images on a TV screen of the sun (1.4 million kilometres in diameter) and a virus (100 nanometres in diameter)—both spherical and filling the screen—without fully realizing that they are separated in size by some sixteen orders of magnitude.

For serious microscopy, however (don't forget that Leeuwenhoek was achieving magnifications of 266 times), coming to terms with the actual scales involved can be a little difficult. Let us start with a familiar item, and progressively magnify it, as one might during observation with a light microscope. Most definitive postage stamps measure roughly one inch (2.5 cm) by three-quarters of an inch (2.0 cm). At a magnification of times ten this stamp becomes roughly the size of a standard A4 sheet of paper, and its area (length multiplied by breadth) has increased to contain one hundred stamps). Increase this by a further ten times to a magnification of one hundred, and the stamp is now the size of a double bed sheet (2.5 by 2 metres, with an area of ten thousand stamps). A further ten times magnification (to one thousand) increases our stamp up to the size of a basketball court (25 by 20 metres, roughly with an area of one million stamps). At this point (times one thousand), we have just reached the top of the useful range of magnification for light microscopy. In terms of perception, most of us would probably be comfortable with the idea of the area of ten thousand stamps (one hundred in length and width), but beyond this it becomes increasingly difficult, bordering on mind-boggling, to envisage such large numbers of a small item. If we proceed to the levels of magnification used in electron microscopy, which can go up to magnifications of one hundred thousand to one million, then the size of our stamp at a magnification of one million would be 15 miles by 12 miles

(25 by 20 kilometres). The area (one million stamps by one million stamps) would be filled by a thousand billion (i.e. a trillion) stamps, certainly an unfeasibly large number to imagine. Perhaps it is simpler to start with something much smaller, but still commonplace, such as a human hair, which has a diameter of around one tenth of a millimetre. Magnification by a factor of one thousand results in a hair ten centimetres in diameter, something we could grasp with both hands. However, to perceive a hair ten metres in diameter (at a magnification of one million) becomes difficult again, but these are the magnifications required to visualize molecules and atoms.

The units of microscopic measurement

The most common units used for microscopic dimensions are micrometres (μm) or microns(μ). A micrometre is one thousandth of a millimetre, or one millionth of the SI unit of length the metre (metre x 10^{-6}). The micron was officially replaced as an SI unit by the micrometre in 1967, but still remains in regular use. Each micrometre is made up of a thousand nanometres (nm) (or one metre x 10^{-9}). One tenth of a nanometre is an angstrom, named after the 19th century Swedish physicist Anders Angstrom—also not strictly an 'official' SI unit these days, but persisting in regular use, largely because it is extremely useful in denoting the average atomic diameter. Each nanometre is made up of one thousand picometres (pm) (or one metre x 10^{-12}), so an angstrom is equivalent to 100 picometres. To give an indication of actual sizes in the natural world, viruses are around 30 to 300 nanometres, bacteria are around a micrometre, whole cells are around 15 to 30 micrometres (although certain neurons are a couple of metres), and parts of cells such as chromosomes are roughly a micron in diameter by 5 to 10 micrometres in length (Figure 3). In the materials world it is a little harder to come up with familiar man-made examples in this size range, although manufacturing processes in nanotechnology can produce gaps in microcircuits measured at 2 nanometres, and functional MEMS (micro-electro

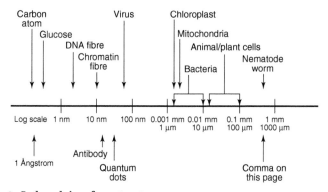

3. Scale and size—from atom to worm.

mechanical systems—tiny components which usually contain sensors and a processing unit) can be manufactured at dimensions of 20 to 200 micrometres.

The basic parameters of microscopy—illumination, contrast and resolution

Although visible light was the only source of illumination that was available to the original microscopists, we now need to consider the whole of the electromagnetic spectrum (Figure 4). The parts of the spectrum which have wavelengths longer than light such as radio and TV are too long to be of use in microscopy, although microwaves have been used in scanning and hybrid microscopes. Electromagnetic energy is easily converted to acoustic energy, and used for imaging, most familiar in ultrasonic imaging in medicine. Scanning acoustic microscopy uses focused sound waves, producing useful resolution in the testing of materials and as quality control in manufacturing of microcircuitry. Within the realm of visible light, blue and ultraviolet light have wavelengths roughly half that of red light, so an improvement of twice the resolution is available at this end of the visible spectrum. X-rays have a smaller wavelength still, but, despite their highly desirable potential for microscopy due to their penetrating power, have

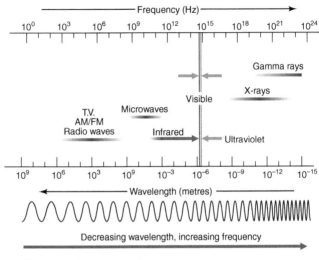

Frequency (Hz)

10^0 10^3 10^6 10^9 10^{12} 10^{15} 10^{18} 10^{21} 10^{24}

Gamma rays

X-rays

Visible

Microwaves

T.V.
AM/FM
Radio waves

Infrared

Ultraviolet

10^9 10^6 10^3 10^9 10^{-3} 10^{-6} 10^{-9} 10^{-12} 10^{-15}

Wavelength (metres)

Decreasing wavelength, increasing frequency

4. Diagram of electromagnetic spectrum.

Microscopy

proved to be fiendishly difficult to control until recently
(Chapter 7). Lenses for X-ray microscopy are made from
concentric rings of gold or nickel on a silicon dioxide substrate,
called Fresnel zone plates. X-ray microscopes can produce
resolution of around 10 nanometres, and X-rays have the
advantage that they penetrate most materials, but originally
required a full-blown synchrotron as a source, whereas electrons
can be generated 'merely' by heating a tungsten wire to dull redness.
Electrons (originally called cathode rays) have a short wavelength,
producing a thousand-fold improvement in resolution down to 0.1
of a nanometre (our old friend 1 angstrom). On the down side, their
lack of penetration requires intensive specimen preparation for
transmission electron microscopy, but is an advantage for scanning
electron microscopy (where they 'bounce off' surfaces).

We must briefly consider how light and electrons behave in order
to understand how magnified images are formed. When light
strikes an object, it may be reflected (i.e. the return of incident

20

light at an interface such as a mirror), or if the object is transparent (water or glass) then light undergoes a change in direction according to the refractive index of the medium. Refractive indices are calculated from the speed that light moves through a particular medium in comparison to the speed of light in a vacuum (which has a refractive index of 1). The change of direction is caused by the slowing down of light as it passes through, and this is determined by the density of the medium. The refractive medium of air is 1.003, water is 1.33, and quartz glass 1.54. As the refractive index increases, so does the property to bend light rays, which is important in lens manufacture.

Isaac Newton knew that light travelled in straight lines and, because of this, likened it to a beam of particles. By the end of the 18th century, however, it was generally conceded that light behaved as a wave largely as a result of elegant experiments performed by Thomas Young in 1803, showing how light spreads out from a slit. Because light is a wave, it bends around corners, and it is scattered when leaving an object. If the peaks of the scattered waves line up, constructive interference occurs to produce brightness. If some of the peaks coincide with the troughs, destructive interference occurs, producing darkness (Figure 5); this is easily demonstrated by viewing a street light through a fine regular weave such as the fabric of an umbrella, which produces a diffraction pattern of bright spots. Similarly, the passage of electrons through a crystalline object produces a pattern of spots which are characteristic of the crystal itself. None of this, however, explains the photoelectric effect, where light exerts a pressure and therefore must be particulate. For simplicity and the brevity necessary for this book, light is made of photons, which have the properties of both a wave and a particle.

Contrast formation

In Julian Heath's excellent *Dictionary of Microscopy* (2005), contrast is defined as 'the difference in absolute or perceived

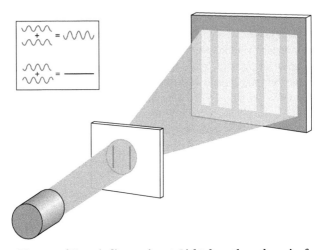

5. Diagram of Young's slit experiment. Light shone through a pair of slits produces several light and dark bands, according to whether or not the light waves (which have diffracted, i.e. bent, as they exited the slit) now have their peaks together, producing constructive interference (seen as bright bands), or peaks and troughs lined up, producing destructive interference (seen as dark bands), as shown in inset.

intensity between an object and its surroundings. Contrast may be physiological or photometric.' More simply, regardless of magnification and resolution, without contrast, there will be nothing to see. The impressive performance of the human eye, which allows us to see a golf ball hundreds of metres distant, is entirely dependent on the contrast between white ball and green background, as a green ball would be virtually invisible. Contrast in microscopy comes from interactions between the illuminating beam (e.g. white light, UV, microwaves, sound waves, X-rays, or electrons) and the specimen. These interactions depend on whether or not the illuminating beam passes through a thin specimen, or interacts with the surface of a bulk specimen. For thin specimens, the illuminating beam passes through and generates a transmitted image, which is standard for conventional light microscopes and transmission electron microscopes. For bulk specimens, interactions occur at the surface, such as in the scanning electron microscope, where the electron beam is focused to a small spot and scanned to and fro over the specimen (in a 'raster'). The interactions between beam and specimen are then detected by a sensor, and processed to form an image on a monitor. Scanning specimens with an illuminating beam is also regularly used in light microscopy to form a transmitted image, as we shall see in Chapter 3.

The role of specimen preparation

Living cells are largely transparent. Without the new dyes that were being produced by the emerging chemical industry at the end of the 19th century, early microbiologists would not have been able to discover and identify the bacteria which were the causative factors for the mass diseases of the time. Paul Ehrlich was one of the first to realize that organic chemicals (dyes) could react with microbes and tissues. Those dyes which had a specific ability to interact with and kill bacteria provided the first steps in the production of antibiotics. In 1910, Ehrlich produced Salvarsan, the first effective treatment for syphilis, a drug that was the first of a series of compounds called sulphonamides which were to revolutionize medicine. Ehrlich also used staining reactions to classify the different types of blood cells and was the first to show the differences between lymphocytic and myeloid leukemic cells, enabling an early classification of two types of leukaemia.

The ability to differentially stain biological material has given us the disciplines of histology (the study of the structure of cells and tissues) and histochemistry (which deals with their chemical composition). Staining at its simplest requires a thin specimen such as a blood smear on a glass slide which is chemically stabilized (or 'fixed'), in some way, often by no more than a quick dip in acidified alcohol (which also dehydrates the specimen), followed by staining and mounting under a coverslip in a suitable mounting medium such as Canada balsam (a mixture of resins from the balsam fir), which dries as a transparent film. Slides made in this way are stable for at least a century, although these days Canada balsam has largely been replaced by synthetic resin alternatives.

To penetrate tissue with light, it must be sliced into thin sections, then stained and mounted. Rigid tissues from plants can be sectioned freehand with a razor blade, but soft material needs to be chemically stabilized (fixed), then supported by infiltration with an embedding medium such as paraffin wax (as used for centuries in light microscopy, or LM). Sections for electron microscopy need to be much thinner yet retain resistance to the heating effects of the electron beam, so they are infiltrated with resins such as araldite, and then polymerized. Wax sections are cut on a microtome (from the Greek *micros*, 'small', and *temnein*, 'to cut'), first produced around the end of the 18th century. Wax sections for LM are usually cut with tungsten blades, whereas the ultramicrotomes used to cut ultrathin resin sections for electron microscopy use knives made from broken glass, or specially machined diamonds.

An alternative preparation strategy, which can reduce the changes that occur during conventional preparation (i.e. fixation, dehydration, staining, infiltration, and embedding), is to freeze small pieces of tissue, which are then sectioned in a cryotome. This procedure is also sufficiently rapid to allow samples from a patient on the operating table to be examined by a pathologist, enabling him to advise the surgeon while the patient remains on the operating table. Freezing specimens is also used for electron microscopy requiring hardware that uses a blast of liquid nitrogen (at –196 degrees C, delivered at high pressure) for rapid freezing, together with low-temperature specimen stages in the microscope so that the sections remain frozen (unlike LM).

Histological staining in light microscopy employs coloured dyes which provide contrast by their different reactions within the tissue, whereas histochemical stains are used to identify individual elements. In biological tissue, virtually any molecule can be identified by raising antibodies to them. The antibodies are then attached to recognizable probes, which are often fluorescent, a technique termed immunohistochemistry. For electron microscopy (EM), where images are always monochrome, resin sections are stained with solutions of heavy metal salts such as uranium acetate and lead citrate, providing contrast by absorbing the electron beam. Immunohistochemistry probes in EM are usually attached to tiny particles of colloidal gold, which are electron opaque, and show up as black dots on the image.

Resolution

As we noted earlier, without an increase in resolution, all the magnification in the world is useless, as things will just get bigger and bigger without any more detail appearing. A good example is magnification of newsprint—which quickly reaches a point where it becomes apparent that it is formed of individual dots. Once the dots have been seen, magnifying them to the size of footballs produces no more information. This is known as empty magnification, and is seen in digital images where individual pixels become apparent (pixelation). For effective or useful magnification, we must continue to see increased detail with each step of magnification. This means that features will continue to be separated as individual entities until the limits of resolution are reached, at which stage two points in the image are no longer separated and appear as one. The minimal distance at which two points appear as discrete items is termed the optimal resolution, or resolving power (around 200 nanometres for a conventional light microscope, or 1 angstrom for transmission electron microscopy).

Resolution is defined by Abbe's equation, which states that resolving power equals half the wavelength used (light in his case) divided by the numerical aperture of the objective lens. According to the laws of physics this holds good for conventional transmitted images in either light or electron microscopy. As the wavelengths of light or electrons are constant, the limiting factor becomes the numerical aperture of the lens used. Numerical aperture is a dimensionless measure of the light gathering and resolving power of a lens. Mathematically (and the only equation in the book),

$$\text{Numerical Aperture} = n \sin\theta$$

where n is the refractive index of the medium in which the lens is working (1.0 for air, 1.33 for water, up to 1.56 for oils), and θ is half the angle of the maximum cone of light (or electrons) that can enter or exit the lens. Because of the big difference between the refractive indices of air and oil, it is clearly beneficial to use a drop of oil between the coverslip of the specimen and the objective lens—hence 'oil immersion lenses'. For dry lenses the maximum NA is around 0.95, and for oil immersion lenses (with a magnification of 100 x) the NA is around 1.4. Even with oil immersion however, the wavelength of light is still limiting, so although a lens with a magnification in excess of x 100 could be built, there would be little point as the resolution would not be improved. However, in 2012 Olympus produced a lens for TIRF (total internal reflection fluorescence) microscopy, using a special sapphire glass and immersion oil with a high refractive index which has a numerical aperture of 1.7—the highest so far. Relatively recently, after centuries of the acceptance of x 100 as the highest useful magnification, manufacturers are offering 150 x objectives in specialized applications. There are several other factors which limit resolution, which we will discuss in Chapter 3.

Chapter 3
Light microscopy—from Abbe to superresolution

Although it is generally accepted that Leeuwenhoek's contribution to microscopy was immense, his single lens instruments left little scope for development. Consequently, Hooke's compound microscope, which used two lenses, provided the starting point from which microscopy has developed to this day. A compound microscope has an objective lens close to the specimen which produces a magnified intermediate image. This image is then further magnified by an ocular (Latin *oculus*, the eye) or eyepiece lens to produce a virtual image at around 25 cm from the eye (Figure 6). A standard modern instrument will have eyepieces (with ten times magnification), and usually a number of objective lenses with magnifications between ten and a hundred. The objective lenses are mounted on a revolving nosepiece, enabling successively higher power objectives to be clicked into place, with no loss of focus. Final magnification is the product of objective and eyepiece magnifications, e.g. 100 (objective) times 10 (eyepiece) = 1,000 times magnification. Improvements in lens design together with computerized production techniques have considerably improved the original 'best magnification' of one thousand, and, as we shall see later, the limits of resolution imposed by Abbe's equation have also

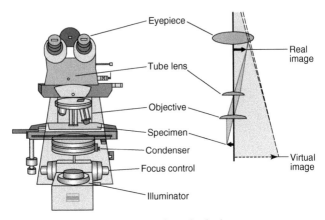

Eyepiece

Real image

Tube lens

Objective

Specimen

Condenser

Virtual image

Focus control

Illuminator

6. Light path diagram and image of standard microscope.

been circumvented; but first let us look at the rich variety of techniques now used across the life and material sciences.

Microscopy

Light sources

In the early days of microscopy, light sources were limited to either the flame of a candle or the sun. While the sun has many desirable features, such as a particularly high brightness level, its use in day-to-day microscopy (particularly in northern Europe) is not the most reliable proposition. Standard light sources these days are tungsten halogen lamps (around 100 watts), mercury and xenon arc lamps, metal halide sources, and, more recently, LEDs (light emitting diodes). Tungsten halogen sources are the workhorses for the majority of microscopy, with mercury and xenon used for fluorescence microscopy. Lasers supply light with two useful characteristics (single wavelength and coherent light), and their early limitations of bulky size and high cost are much reduced by progressive development. One of the defining features of modern microscopy, as we shall see throughout this volume, is the speed with which technological developments in related fields become incorporated.

Bright field microscopy

Conventional bright field microscopy is performed in transmission mode, in which the specimen is illuminated from below with light passing to an objective lens and an eyepiece lens. There may also be an intermediate tube lens in the production of the final magnified image. From the origin of the microscope illumination (either a mirror deflecting a light source, or a lamp system built into the base of the microscope) the light passes through the condenser, which is mounted below the specimen stage. The condenser functions to gather light and concentrate it into a beam that illuminates the specimen with uniform intensity over the entire field of view, with the intensity controlled by a variable iris diaphragm (essentially a variable aperture much like that on a camera). Microscopes with integral light sources also have a field diaphragm. Way back in 1893, in the Zeiss factory, August Köhler (1868–1948) introduced a method for optimal illumination of the specimen which remains the standard method to this day. The first thing for any aspiring microscopist when sitting down at a microscope is to check for Köhler illumination. It is a straightforward procedure, requiring no more than a few simple adjustments (adjusting the centring of the light filament, field diaphragm, and condenser aperture diaphragm) to guarantee that the specimen is evenly illuminated, but often (sadly) neglected in practice.

For specimens that the beam cannot pass through, and also regularly for fluorescence microscopy, the specimen is illuminated from above, by a series of prisms which route light from a second source (an epi-illuminator) mounted halfway up the back of the microscope column. The light is directed by a beam splitter down through the objective, and light reflected by the specimen through the objective and beam splitter forms the intermediate image. This mode is called reflected light microscopy, and is a standard method for material sciences where specimens are usually opaque, although for example rock samples may be polished to sufficiently thin slices for light transmission.

Objective lenses

Once the illumination is set up, the standard microscope requires only the addition of a suitable sample, which is generally mounted on a 3 inches by 1 inch glass slide and covered with a thin glass coverslip. Light passing through the specimen then reaches the front lens element of the objective lens. Modern objective lenses contain multiple optical elements and a range of magnification between x 2.5 and x 150. Good quality lenses are manufactured to overcome spherical aberration (where different rays are brought to a different point of focus, degrading the image), and are also compensated for at least two colours of chromatic aberration (preventing coloured haloes around the subject due to the different wavelengths of each colour). Lenses compensated for two wavelengths are called achromatic, or apochromatic for compensation for three or four different wavelengths. Lenses may also be adjusted to ensure that the whole specimen image is in focus in the same plane to produce a flat field of view (or plan lenses). Each correction requires the addition of more elements within the lens, with an increase in cost, so that a planapochromatic lens will be considerably more expensive than a simple achromatic lens.

Most lenses below x 60 magnification will be 'dry', meaning that the light travelling from the specimen will be refracted twice, once as it leaves the coverslip and secondly as it enters the front element of the lens. In order to adjust for this, lenses of x 60 and above will have this gap filled with a drop of 'immersion' oil, which has a refractive index to match that of the coverslip. Oil immersion is standard, but water immersion is available, and often used without a coverslip for living aqueous organisms. Immersion lenses will be spring loaded to protect the front element (if it gets driven through the coverslip the consequences are disastrous!). Immersion lenses may also have a correction collar which can be rotated to adjust for variations in coverslip thickness. With the incorporation of increasing numbers of lens elements, the diameter of the lens increases to a point where it can be wider than its height, requiring a specialized nosepiece or turret (the rotating part into which various objectives are screwed). Most immersion lenses have a magnification of 100 x, although higher power lenses (at x 150) are a relatively recent innovation, and reflect the role of computer aided design not only in design, but also in manufacture, so that the skill of the lens grinder has been replaced by the precise consistency of computer controlled machinery. Different types of glass, having highly specific refractive indices, are also incorporated, yielding images that are remarkably crisp, bright, and sharp, and way beyond the wildest dreams of early microscopists. The single most important parameter of an objective lens is its numerical aperture, which is, as we have seen, the measure of its light gathering ability and resolving power. The theoretical limit of numerical aperture for oil immersion lenses is a value of 1.51, but in practice most good lenses have numerical apertures of around 1.0 to 1.3, and at the top of the range, 1.4 (with a suitably increased price level).

Eyepiece lenses

As the resolution of the image has already been determined by the objective, the function of eyepiece lenses is to magnify the intermediate image (the second stage in a two stage imaging process). In order that the observer may use both eyes in a binocular microscope a prism (beam splitter) splits the image and directs it to both eyepieces. Eyepiece magnifications vary between x 2.5 and x 25 (but high eyepiece magnification will not increase resolution, and merely produce 'empty magnification'). Eyepieces have a front eyepiece, and additional multi-element lens, and may be 'widefield' for a wider field of view, or 'high eyepoint', where the image is formed at a height of around 25 mm from the lens, so that it can be used without the need to remove spectacles. Eyepieces can also incorporate graticules for measuring, or grids for counting. One of the earlier ways of image recording was a 'camera lucida' in which the eyepiece lens was incorporated into a combination of lenses and prisms that produced a projected image onto a piece of paper beside the microscope as an aid to drawing. The wide availability of a whole host of cameras these days means that many users will work directly from a TV monitor for their image viewing rather than actually 'looking down' the microscope. As we shall discuss later (Chapter 7), much of 'conventional' microscopy, particularly in terms of inspection and quality control, is able to be remotely controlled so that the observer will only view an image when alerted on account of some irregularity perceived by a computerized image analysis. Computer controlled motorized specimen stages can also be set to seek out a particular feature of a specimen and record its position, so that the operator can view it directly without time spent searching. This is particularly useful in human chromosome analysis, where even in the best preparations, it can be extremely time consuming to find a good chromosome spread.

Recording microscopical information

For a period of almost fifty years from 1680 onwards, Antoni van Leeuwenhoek sent letters to the Royal Society of London, with detailed drawings of almost everything he viewed in his microscopes, including sand grains, human teeth and muscles, and various types of wood. A typical example was an exquisite drawing of duckweed, along with the animals which lived attached to it, sent in 1702. The drawing (still retained in the Society archives and accessible on their website) was in red chalk on paper, measuring just less than A5. This drawing showed ciliates, vorticellids, and rotifers ('animalcules') along with the budding of Hydra, all in remarkable detail. Drawings remained the only way of microscope recording until the dawn of photography, but were an art form in themselves, reflecting the long hours of observation in their production and an amazing grasp of detail.

Photography was in its infancy at the start of the 19th century, and became reasonably widespread by 1850. Early microscopists were aware of the possibilities of photographic recording, with both Pasteur and Koch pioneering the use in recording images from the microscope, which were limited by the lack of detail due to the stage of development of photographic emulsion. Mounting a plate camera on a microscope itself was not a big problem, and this became the standard for microscopic photography for decades, with improvements dependent on advances in photographic emulsion. Plate cameras were replaced by 35 mm roll film, mostly in a camera back mounted on top of the binocular viewing head, with the exception of the Zeiss photomicroscope (*c*.1950), where the camera was neatly inserted into the body of the microscope. After the arrival of colour, microscopists would eagerly await the introduction of more advanced emulsions, particularly in fluorescence microscopy, where the challenge was to expose the film before the fluorescence faded, leading to dreaded 'reciprocity failure' (in which the film becomes less sensitive as the length of exposure is increased). This created a vicious circle driving microscopists back to monochrome film (which was much faster, but obviously unable to show different colours).

The advent of cine film also opened the way for time lapse microphotography, which for the first time allowed the general appreciation of events such as cell division, which takes around an hour in real time, with changes occurring at a barely perceptible rate. Once speeded up by around 200 times, however (one frame exposed every ten seconds and played back at 25 frames per second), the full beauty of the event can be appreciated over a few seconds.

With the advent of video tape recording, a video camera could be set up on a microscope in a temperature controlled environment, allowing continuous time lapse observation of cellular interaction. Having set this up in my own lab, we would record cultures of bone marrow cells for 24 hours a day, and weeks at a time (using a one hour tape cassette over the course of two weeks). In this way we were able to watch the development of red blood cells through several rounds of division, culminating with the expulsion of the nucleus in the final stage of differentiation.

The development of digital imaging, which combines modern optics and high resolution imaging chips, has led to a whole new area of microscopy, namely digital microscopy, which we will review later (Chapter 6). For the rest of this chapter we will survey the different types of light microscopy that have resulted from novel ways of manipulating light, and the incorporation of new sources of illumination.

Dark field and phase contrast microscopy

As most living material is largely transparent, chemical preservation (fixation), followed by sectioning and staining, was the main method of generating contrast for early microscopists. This obviously precluded examination of living material, although some improvement in contrast was produced by dark field microscopy, where the light is focused from the condenser so that only scattered light from the specimen is imaged by the objective lens. The invention of phase contrast microscopy by Frits Zernike enabled the first visualization of internal details of living cells and bacteria, and earned him the Nobel Prize for Physics in 1953. Zernike's breakthrough was based on his studies of optical diffraction in the 1920s. As the light passes through cells, it slows down slightly, producing a phase shift relative to the surrounding light. The phase shift is invisible to the human eye. If, however, this change is increased to half a wavelength by the use of a transparent phase plate, it will destructively interfere with the light that has not passed through the specimen, and produce a significant difference in contrast (Figure 2b). The optical system requires two phase plates, in the form of annuli, one in the condenser (beneath the specimen) and another in the objective lens itself. The phase plates are produced by etching of the glass surface, which produces a phase change without affecting the optics. Zernike's discovery was not initially welcomed at the Zeiss factory, where (as recorded in his Nobel acceptance speech) 'there was a strong feeling that everything useful in microscopy had already been invented'.

Polarized light microscopy

Just as the phase contrast microscope takes advantage of the phenomenon of interference in producing contrast, the polarizing microscope uses plane polarized illumination (i.e. light which has been passed through a polarizer, producing light which only oscillates in a single plane). Some types of specimen have properties which modulate polarized illumination. 'Anisotropic' materials display optical properties which differ according to the direction of polarized light illumination. This type of specimen is usually one in which there is a high degree of molecular order, such as a crystal, which is observed in the polarizing microscope with a second polarizing filter (the analyser) positioned at the objective lens. The analyser is rotated to block all the polarized light from the condenser, so that any variation in polarization produced by the specimen will show up as an alteration in contrast, or often (spectacularly) as colour changes. Polarizing microscopy produces images which rely on interference between so-called ordinary and extraordinary rays from anisotropic material. Colours produced in this way can identify unknown parts of a specimen (such as a thin slice of rock) by reference to a chart produced by Michel Levy, a French geologist, around the turn of the 20th century. As well as a standard method in mineralogy in the identification of rock types, polarizing microscopy is also used in biology, where hard tissue such as tooth and bone display anisotropic properties, and even in plant studies, where there is a high degree of molecular orientation in the cellulose fibrils of the cell wall. There seems to be no particular inventor of polarizing microscopy, although Dr Edwin H. Land, who invented polarizing film in 1932, should take some of the credit. Polarizing microscopes also need to be optimized so that none of the components in the light path interfere with the plane of polarization, and are equipped with rotating specimen stages and nosepieces to centre the rotation to the field of view. As with other microscopes, they also can be used with reflected light as well as transmitted light, so that thick or opaque specimens, for instance silicon wafers, can be examined and analysed.

Differential interference contrast (Nomarski) microscopy

Whereas phase contrast microscopy is not known by the name of its inventor, the name of Georges Nomarski is still commonly used to describe his differential interference contrast (DIC) system for enhancing contrast in transparent specimens. DIC incorporates polarized monochromatic (usually green) light, which then passes through a Wollaston prism below the condenser before striking the specimen. The Wollaston prism splits the light into two beams separated by a 'shear' distance. The two beams then enter the specimen, where they are altered by the different parts of the specimen, and subsequently recombined by a second Wollaston prism, together with a second polarizer (analyser) set for maximum extinction of the background illumination. DIC produces an image with both brightness and colour changes, and a three dimensional appearance (pseudo-relief) similar to the images from scanning electron microscopes. This appearance is not truly representative of the actual specimen geometry, and although both attractive and informative, cannot be used for any measurements of height. The advantage of DIC is that no phase plates are required (as for phase contrast), and a thin plane of the specimen is focused, which allows optical sectioning at different heights within the depth of the specimen. Differential interference contrast is equally popular with phase contrast amongst biologists, and research microscopes will often be set up for either, depending on the particular specimen or requirements of the operator.

A similar imaging mode to DIC, but with a greater range of induced colour in the specimen, was produced by Dr Robert Hoffman in 1975 (Hoffman modulation contrast). In his system two polarizers are inserted below the specimen, together with a modulator plate placed in the objective lens. This arrangement produces a combination of both phase contrast and DIC image, generated by variations in the specimen.

So far we have considered the effects of varying the type of illumination, so at this point we can sum up how one specimen can be imaged in four separate ways. In a conventional microscope with bright field illumination, contrast comes from absorbance of light by the sample (Figure 7a). Using dark field illumination, contrast is generated by light scattered from the sample (Figure 7b). In phase contrast, interference between different path lengths produces contrast (Figure 7c), and in polarizing microscopy it is the rotation of polarized light produced by the specimen between polarizer and analyser (Figure 7d). This is 'converted' into an image that has colour and a three dimensional appearance by the use of Wollaston prisms in differential interference microscopy. For virtually any specimen, hard or soft, isotropic or anisotropic, organic or inorganic, biological, metallurgical, or manufactured, there will be a variety of imaging modes that will produce complementary information. Some of the types of light microscopy we have looked at above have direct parallels in electron microscopy (Chapter 4).

Imaging outside the visible spectrum

While the early microscopists were ingenious in their manipulations of visible light, the appearance of sources of ultraviolet and infrared in microscope illumination was a natural consequence of the incorporation of any novel optical advance into the realms of microscopy. Because UV has a shorter wavelength (around 300 nm) than the 400–800 nm of visible light, Abbe's resolution limit (dependent on the wavelength) can be reduced to 0.1 nm using a UV source. Using blue and UV light would seem to be a good option, except that the human eye is not sensitive to UV (but easily damaged by it), and glass is not transparent to wavelengths below 380 nm. Both of these problems can be circumvented by the use of quartz lenses and photographic recording. Köhler himself, experimenting with blue and UV light in 1904, noticed that certain tissues fluoresced when illuminated with UV, and thus 'discovered' fluorescence microscopy, describing auto-fluorescence, although

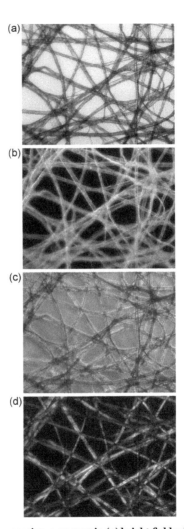

7. Image of a paper tissue, as seen in (a) bright field microscopy,
(b) dark field microscopy, (c) phase contrast microscopy, and
(d) polarizing microscopy.

he saw it more as a nuisance than a useful characteristic. It was to take another forty years before suitable fluorescent probes became specific enough for fluorescence microscopy to take off.

UV microscopy without probes is still useful in biology (when viewed indirectly with video imaging to prevent damage to the eye). Both proteins and nucleic acids absorb UV allowing visualization detail in the cell nucleus and other cellular components. Bundles of a fibrous protein called actin, a component of the structural support within our cells, are difficult to visualize in electron microscopy due to lack of contrast, but can be clearly defined using UV microscopy, particularly in time lapse studies. UV microscopy is also useful in monitoring the growth of protein crystals, and also as an indication of the purity of the constituents in pharmaceutical design and production. Crystals are induced to form in a droplet of protein-saline solution, in a vapour diffusion method that also leads to the formation of salt crystals. Both salt and protein crystals appear identical under a conventional microscope, but salt crystals do not absorb UV, whereas protein does, dramatically separating them. Producing crystals of purified protein allows molecular structural organization to be investigated by electron spectroscopy, and for difficult compounds, attempts may take several years. UV microscopy is also used to show up contamination of microcircuitry that cannot be seen otherwise; to identify specific dyes in textiles; and expose underlying layers in works of art. The light sources of modern UV microscopes also have a strong output in the near infrared (NIR). One of their more important uses for NIR is in imaging microcircuitry embedded within silicon wafers. Conventional imaging only shows surface details, but in the NIR region of the spectrum, silicon is transparent, allowing internal devices to be imaged directly.

Fluorescence microscopy

The ability of certain minerals and extracts of plants to glow in the dark (phosphorescence) was first recorded in the 16th century.

A Spanish botanist, Nicholas Monardes, described that a yellow aqueous extract of kidney wood had a distinct blue tint in reflected light. Three centuries later, in 1852, George Stokes found that feldspar (calcium fluorite CaF) would emit red light when 'excited' with blue light. He named the phenomenon fluorescence, and noted that the fluorescent emission always had a longer wavelength than the exciting light (Stokes shift). Fluorescence occurs as a result of the energy from absorbed photons being re-emitted at a longer wavelength when electrons are momentarily excited (which lasts for about one hundred millionth of a second) and then return to their original ground state.

A fluorescence microscope is based on a standard light microscope, equipped with a source of excitation light—a mercury or xenon arc lamp (or laser)—and filters that are specific for excitation and emission wavelengths; if red fluorescence is the result of blue light excitation, then all but red light needs to be filtered out of the image. Fluorescence microscopy came about originally from attempts to use the shorter wavelength of blue and UV light to improve resolution by Köhler in 1904. Although UV light is invisible to the human eye, it was first shown to exist in 1801 by Johann Ritter in studies of the chemical changes of silver chloride induced by UV light. Ordinary glass is not very transparent to wavelengths below 380 nm, so that lenses for UV were made from quartz.

Some fifty years later, Albert Coons and Thomas Weller (a Nobel prize winner for the growth of polio virus in 1954) developed a staining method for polio virus. The presence of the virus could be identified on cells in tissue culture, using a primary antibody specific for one of the viral proteins, which in turn was labelled by a secondary antibody tagged with a fluorescent marker (fluorescein). For the first time, the presence of a virus (much too small to be resolved by light microscopy) could be identified by a fluorescent signal, a breakthrough of massive significance in cell biology, molecular biology, immunology, and medicine. This advance allowed any large molecule to be tagged with a fluorescent probe providing there was a suitable antibody. As well as fluorescence microscopy, techniques such as flow cytometry (cell sorting), and methods for DNA analysis and gene array, are all based on the use of fluorescently labelled probes. At this point we shall leave fluorescence microscopy (returning in Chapter 4), to introduce the next major advance in light microscopy, in which the light is focused to a small spot which scans over the specimen in a raster, rather than the previous method of flooding the entire area with illumination, now known as widefield microscopy.

The beginnings of confocal microscopy

Widefield microscopy has limitations that arise from illuminating the entire specimen which causes multiple light scattering events, detracting from the quality of the image. Sharp focus is limited to a small portion of the specimen depth (even if it is a relatively thin slice of material), so that the rest of the specimen depth contributes only out of focus information to the final image. However, if the specimen is imaged with a tiny spot of focused light, then the contribution of light scattering and of out of focus information in the image is considerably reduced. If a pinhole is positioned in the image pathway, the out of focus content can be blocked altogether. This is called a confocal image, with considerable improvement over the conventional 'widefield' image. However, one tiny point of illumination is not going to generate much information in respect of the whole specimen, so an image must be built up point by point, and line by line, in a raster arrangement (much like the formation of a television image). Building an image in this way was solved initially by Marvin Minsky, at Harvard, who kept the tiny spot of light stationary, and moved the specimen in a raster, building an image over time. This arrangement did not really catch on, however, although Minsky had the foresight to patent the concept in 1957. Confocal microscopy did not really advance as a working practice (Figure 13) until Brad Amos and John White produced a working prototype in Cambridge in the mid 1980s (Figure 8).

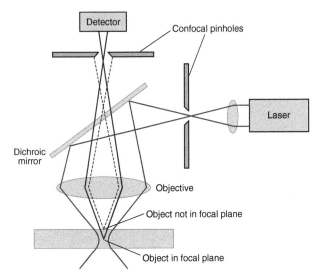

8. **Light path diagram of confocal microscope.**

Spinning discs

Some ten years after Minsky's patent, Mojmir Petran and colleagues at the Charles University in Prague adopted a different approach, using a pair of Nipkow discs (one to illuminate the specimen, the other to collect the image). Nipkow discs have numerous tiny holes (around a thousand) arranged in a spiral pattern, which produce a scanning pattern with multiple points of light, meaning that the whole image is produced in real time (rather than building up a rastered image). This was an arrangement used in early stages of television development many years earlier by Nipkow himself (1884). Larger holes meant more light was transmitted, creating a brighter image, whereas smaller holes made for better confocal images. However, only a very small proportion of the illumination gets through to form the image (around 4 per cent), somewhat limiting this technology, although spinning disc confocal microscopes were produced commercially (after the first few

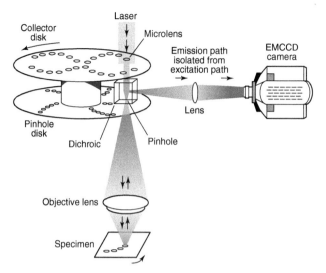

9. Light path diagram for spinning disc microscope.

prototypes were built in a Prague brewery). The next step, however, made by the Yokogawa corporation in Japan, was to insert a tiny lens into each of the holes in the Nipkow discs (even with 20,000 pinholes), which allowed more than half of the incident light to be imaged in the system. A further advantage of the spinning disc system is that the illumination and image collection all take place via a system that can be fitted retrospectively to conventional instruments, making use of research grade optics. The speed with which images are formed and collected (up to 2,000 images per second) has made this type of microscopy extremely useful in following dynamic events in living cells (Figure 9).

Laser scanning confocal microscopy

Laser light sources were originally thought of as solutions to questions which had not been asked: they were to be military 'death rays', but only if you hit the pilot of an opposing plane directly in the eye. However, as a bright, monochromatic light source, which did not spread out even between the earth and the moon, they were perfect for confocal microscopy. All that had to be done was to move them over the specimen in a raster, and collect the resultant image. This can be done by moving the spot of light from the laser by a pair of galvanometer controlled mirrors, or by using the properties of certain crystals such as quartz or tellurium which can deflect a beam in a controlled manner under the control of a piezo transducer, which vibrates to compress the crystal and change the refractive index (an acousto-optic beam splitter). This acousto-optical effect can also be used to select individual wavelengths as excitation light for fluorescence (acousto-optic tunable filter).

As well as removing out of focus information from the image, confocal microscopy has literally added a new dimension to microscopy, which is the ability to move up and down through the specimen in the 'z' axis, as opposed to the previous limitation of x and y axes only (side to side and top to bottom).This enables a series of images (optical sections) to be collected at successive levels within the depth of the specimen (called z stacks) which are stored in a computer, and reconstructed as a three dimensional representation. As everything has to be aligned for this process, the confocal microscope will have a precisely controlled mechanical specimen stage, able to move the point of focus through the specimen and collect optical sections at regular intervals. Given this facility, the temptation is to cut much thicker sections, and produce 'deeper' 3D reconstructions, but as usual in microscopy (which can mirror life in general) there is a compromise. As the section increases in depth, it becomes harder to get light through it. This necessitates an increase in the size of the confocal pinhole, which in turn reduces the ability to make sharp optical sections. Even with the thinnest sections, the actual z resolution (sharpness in the vertical axis) is two to three times worse than that in the x and y axes. This is because the light forming the image focused by the lens system is not focused to a single point, but a bright spot called the Airy disc, which is surrounded by a series of concentric bright and dark rings in the x and y plane, and elongated cones of light (in an hourglass shape) in the axial or z dimension, above and below the plane of focus. This spreading of light is called the point spread function (PSF). Eliminating the outer rings increases the overall resolution, but making the pinhole smaller further reduces the very small amount of light being collected to form the image. Because the overall image is being formed spot by spot, line by line, light is collected by a photomultiplier, in much the same way as an image is formed in a TV camera.

Deconvolution microscopy

Deconvolution microscopy is a method in which computerized image processing techniques are used to improve the contrast and resolution of digital images. The dictionary definition is 'a process of resolving something into its constituent elements or removing complication in order to clarify it'. Most users will use deconvolution on images from widefield microscopes (usually fluorescence microscopy). As we have just discussed, widefield images have some out of focus content whereas those acquired by confocal imaging do not (or have a lot less!). What deconvolution does is to either subtract this blurring, or attempt to reassign it to its source. Deconvolution analysis requires 'high computational demand'—to begin with, it took several hours to process a single image. In 2013, real time deconvolution was still some way off, but getting much quicker than before. For thick sections of fixed material such as tissues or whole embryos, where intense illumination cannot damage the specimen (unlike living cells), and 3D reconstruction is required, confocal microscopy is more effective. Since its introduction in 1983, deconvolution microscopy has become a key imaging tool, and the algorithms continue to become more sophisticated, along with the overall advances in computing, which has previously struggled to handle the huge amounts of data generated by the requirements for deconvolution in rapid multidimensional living cell microscopy.

Superresolution microscopy

Light waves undergo diffraction (bending or spreading) as they pass round an edge. As Abbe showed in 1873, this limits resolution to half the wavelength of the light. Consequently, a small emitter of light such as a fluorescently labelled protein molecule with an actual size of 1 nanometre will appear as a spot of light of 200 nanometres in diameter—the diffraction limited resolution of the light microscope. If there was only one molecule, then the actual site of it could be worked out by reducing the spot of 200 nanometres down to its centre. Knowing where one molecule on its own is in isolation, however, is not going to tell us much about its function amongst 100,000 neighbouring molecules. Put more labelled molecules into the same area, and as soon as they are closer than 200 nm, everything merges into a single blurry bright area. The way round this is to limit the number of fluorescent events that are sampled at any one time as the light is scanned over the specimen—still collected in the confocal microscope with the same lenses—but to be able to collect just a few well spread points of fluorescence at any one time, and use them to create accurate localizations of each one. An analogy would be an attempt to localize spots of rain falling on a car windscreen. As soon as enough have fallen to cover the entire windscreen, we have little idea of where each one landed. If, however, we could see where each fell, then we could measure its centre for an accurate location. Turning on the windscreen wipers creates a clear screen momentarily, and if we could photograph the screen in the time when a few spots have just landed, then those points could be mapped and recorded as tiny central spots. Over the course of a few wipes, we could map the precise location of the point of contact of each and every raindrop. We are not actually improving the resolution, but splitting up numerous (interfering) events so that each can be recorded separately, and reduced artificially to a small point, which can now be separated from its nearest neighbour (recorded at a different time). In this way the effective resolution can be improved tenfold (a resolution of 20 nanometres), and, with an optimal specimen such as diamond crystal, down to 8 nanometres. This is an example of what is known as 'breaking the diffraction barrier' or superresolution.

This sampling of a small proportion of bright spots of fluorescence at any one time can be achieved in several ways, which together generated an assortment of acronyms in the first decade of the 21st century such as STORM (stochastical optical reconstruction microscopy), STED (stimulated emission depletion microscopy), PALM (photoactivated localization activation microscopy), FPALM (fluorescence PALM), and PALMFCS (PALM with fluorescence correlated spectroscopy). These techniques are grouped under the term of localization microscopy. It is beyond the remit of this volume to go into detail about each one, but they all work on the same principle of partial sampling of overlapping spots of light generated by fluorescence and collected by confocal scanning microscopy. As we shall see from our considerations of labelling in Chapter 4, this 'breaking of the diffusion barrier' in light microscope resolution has come about by the ability to collect and modify scanned images of fluorescent labels (mainly) in biology, by manipulation of the way that fluorescence from individual molecules is collected. Most have some way of switching the fluorescence on and off at any one instant, and with STED, this is done by scanning the specimen simultaneously with a laser beam configured into a hollow tube, so that an annulus or doughnut shaped beam scans the specimen, suppressing the fluorescence of everything except the fluorescent spots that are at the centre of the doughnut, which are 'allowed' to fluoresce. The smaller the dark area in the centre of the doughnut, then the better the resolution, which is usually around 70 nanometres (around a third of the Abbe limit) (Figure 10). In FPALM microscopy, individual fluorophores are temporarily bleached, then re-energized by a second laser. Much of this progress is dependent on newly developed fluorescent molecules which can be 'lit up' and 'turned off' at will (i.e. photo-activated and deactivated), whereas the original fluorophores used for fluorescence microscopy would

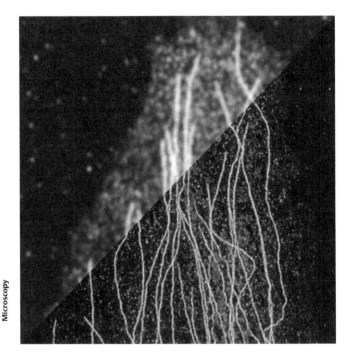

10. Resolution improvement from confocal image using STED.

fluoresce then bleach. Some of these involved illumination and image collection techniques which might appear to require long periods to produce the final image (although STED is in real time), but the rapid scanning capacities of modern confocal instruments can produce images in seconds, allowing dynamic events in living cells to be followed at molecular resolution for the first time, an exciting and significant development just a century and a third after Abbe's original equation on resolution limits. In 2014, Eric Betzig, Stefan Hell, and William Moerner shared the Nobel Prize for Chemistry for their pioneering efforts in this field.

Chapter 4
Identifying what we see

On a dark night, we can see light from a candle up to a mile away, and also from stars hundreds of light years away. Although both are far beyond the resolution of the human eye, we perceive the light (albeit without any real indication of the shape or size of the source) as a bright spot on a black background. This type of image perception is the basis of fluorescence microscopy, in which the light signals are produced by the interaction of the illumination and the specimen, and appear as bright spots on a black background, even though their source may be below the resolution of the objective lens. Even with a modest x 40 lens, we can 'see' fluorescent crystals (quantum dots) which have a diameter of 20 nanometres, around one tenth of standard LM resolution. This is because in fluorescence microscopy the resolution of the lens gives an image at its resolution limit regardless of the actual size of the object, with the same image size seen from fluorescent objects of 20, 100, or 180 nanometres in diameter. Briefly, although the (conventional) fluorescence microscope cannot provide optical spatial resolution below the diffraction limit, ingenious methods have been applied to allow the detection and localization of a single fluorescent molecule. As we have seen briefly from Chapter 3, developments such as STED, STORM, PALM, and FPALM have indeed 'broken' the diffraction barrier in pursuit of better resolution in fluorescence microscopy.

The basics of light emission

When organic or inorganic substances emit light, it occurs as one of three phenomena, phosphorescence, chemiluminescence, or fluorescence. In phosphorescence, inorganic materials such as phosphorus absorb light and then slowly release it over a period of time, producing the 'glow in the dark' familiar from clock and watch dials. In chemiluminescence, light is produced as a result of a chemical reaction, as seen in 'glow sticks'. When similar reactions are produced in living cells, this is termed bioluminescence, which occurs mainly in sea creatures, with a few terrestrial examples in bacteria, and perhaps most spectacularly in fireflies and glow-worms. Fluorescence is a process in which a substance becomes excited by absorbing photons, which results in the emission of a radiation of a lower energy and longer wavelength than the absorbed light, producing a change in colour. As an example, the mineral fluorspar emits red light (longer wavelength) when illuminated by ultraviolet light (shorter wavelength). Unlike phosphorescence, fluorescence emission stops immediately when excitation ceases.

Some constituents of biological systems are naturally fluorescent, as noted by Abbe himself during his studies with the shorter wavelengths of blue light and UV illumination as a route to improved resolution. Naturally occurring fluorescence is termed primary or auto-fluorescence, such as the brick red auto-fluorescence of chlorophyll in plant cells when they are illuminated with UV light. In animal cells, lysosomes (food vacuoles) also show a brick red auto-fluorescence in UV. Auto-fluorescence is also useful in the fields of petrology and semiconductors, but in biology auto-fluorescence is too limited to use for sophisticated analysis. Everything changed, however, when in the 1930s Max Haitinger developed secondary fluorescence, by treating specimens with fluorescent dyes (also called fluorophores), which have the property to absorb light energy, and briefly shift to a higher energy state, which is maintained for only a few nanoseconds before the molecule drops back to its ground state, releasing the absorbed energy as fluorescence.

This cycle is then repeated up to tens of thousands of times, but will eventually fade. Certain fluorophores bind specifically to individual parts of tissues, cells, and various pathogens, causing them to fluoresce, thus lighting up specific areas of the specimen as red, green, or blue against a black background. The source of fluorescence excitation may be a white light source such as a mercury or xenon lamp, which contains several peaks of exciting emission at different wavelengths. Optical filters are used to select the desired wavelengths which are suitable to excite the chosen fluorophores used to stain the specimen. Lasers produce monochromatic (single wavelength) high intensity light, making them ideal for fluorescence microscopy, although more than one may be required, and they can be bulky (and expensive), requiring special systems (optical fibre or liquid light guides) to interface to the microscope optical system. More recently LEDs and laser diodes, which also produce a single wavelength peak, are becoming routine sources, as they are brighter, cost less, and last longer. Most fluorescent microscopy is performed using illumination from above (epi-illumination), where the exciting light travels down to the specimen through the objective lens via a filter cube. The filter cube has a dichroic mirror that reflects light onto the specimen, but also allows fluorescence emission from the specimen to pass upwards to the eyepieces (or camera) (Figure 11). For confocal fluorescence, laser illumination of the specimen also starts with an epi-illumination light path, prior to the single spot of light being scanned over the specimen. Because of the relatively low level of light emitted at any one moment in time as the scan passes over the specimen, the signal is collected by a photomultiplier which builds up the image pixel by pixel.

Dyes (fluorophores) which match their excitation peaks with the wavelengths available from laser sources have also been developed to enable spectacular multicolour staining of several parts of the same cell. Once a particular fluorophore has been excited, a second set of filters between the specimen and the eyepiece ensures that

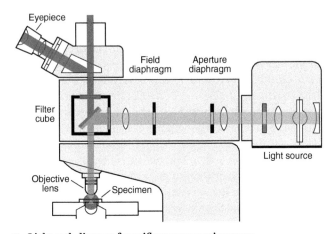

11. Light path diagram for epifluorescence microscope.

only this fluorescent signal will reach the eye of the observer, blocking any stray light which will reduce the signal and contrast. Only a single fluorophore can be viewed at any one time, and the multicolour image is built up by superimposing several signals to make one composite image. Depending on the sophistication (and cost) of the instrument, this is achieved by moving the filters manually, or they may be mechanically driven under computer control. Living cells are photosensitive and, when two or three fluorophores are used at the same time, rapid switching of illumination sources and analysing filters is required to prevent light induced damage (phototoxicity).

Although auto-fluorescence pointed the way for specific identification of individual components of cells, it is too limited to allow any real progress, leading to the development of an entire branch of chemistry devoted to the production of fluorescent staining materials, allowing life scientists to choose from a spectrum of fluorescent probes. These include genetically engineered fluorophores such as green fluorescent protein, fluorescently tagged antibodies for specific proteins, as well as probes for virtually every other cell component such as nucleic acids, and also to visualize the inorganic components of cells such as calcium ions (which are highly significant mediators of biological activity).

The development of fluorescent stains

DAPI (4-6-diamidino-2-phenylindole) is a compound which was originally synthesized in the 1970s as an antibiotic for trypanosomes (which cause sleeping sickness). DAPI binds specifically to the AT pairs of bases in DNA and produces a bright blue fluorescence, and has become a standard fluorophore for the cell nucleus (where the DNA is contained). Because of its specificity for particular sites in DNA ('AT rich regions') DAPI has also revealed a novel substructure in chromosomes (the packaging for DNA at cell division), producing bright bands along the chromosome arms, a discovery in the 1960s that allowed significant progress to be made in chromosome recognition and analysis. Fluorescent staining of DNA in living material has many uses: as an assay for the amount of bacteria in human foodstuffs, in forensic investigations, and as a marker for 'apoptosis' (cell suicide) in both normal and cancer cells. Propidium-iodide also stains DNA, but because it cannot cross the outer membrane of living cells, only nuclei in dead cells are stained, thus providing a useful indicator of viability. Other antibiotics have also been found to be good fluorescent stains for DNA, such as chromomycin, which binds preferentially to the GC base pairs in DNA. These early DNA stains have been developed into a variety of stains for specific sequences of AT and CG pairs in DNA by a process called FISH (fluorescence in-situ hybridization), where fluorescently labelled RNA of a known sequence is used to pick out the complementary sequence amongst the whole DNA of a particular cell, chromosome, or even an invading DNA virus. FISH is so sensitive that it can detect a single copy of a gene (amongst 20,000 others) in an individual cell. Further development using multiple fluorescent probes has produced a technique called chromosome painting which produces a multicoloured staining of human chromosomes, enabling subtle changes and mutations (responsible for genetic disease conditions) to be readily identified.

Other parts of the cell can be directly identified by a variety of fluorophores. Phalloidin, a highly poisonous extract of *Amanita phalloides* (the death cap mushroom), can be chemically complexed (tagged) with a fluorescent molecule, rhodamine, which fluoresces red. Phalloidin binds to a fibrous protein called actin, one of the constituents of the cellular cytoskeleton (Figure 12). Along with DAPI (blue) staining of the nucleus, and tags with different fluorescent colours for other cell constituents, and structures such as mitochondria or microtubule, a multicolour image of the cell contents can be built up in living cells. Any subsequent alterations (natural or experimentally induced) can be observed over time. A recent list of fluorescent dyes used for microscopy numbered well in excess of 200 fluorescent molecules, spread evenly through the entire coloured region of the spectrum from blue (wavelengths

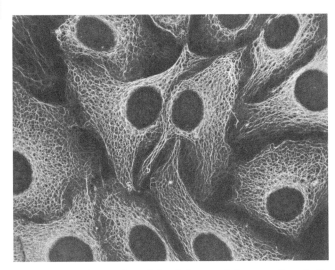

12. **Fluorescent image of cytoskeletal elements in human cultured cell.**

404 to 479 nm) through blue green (480 to 496 nm), green (501 to 538 nm), yellow (540 to 579 nm), orange (580 to 599 nm), red (602 to 674 nm), and finally far(infra) red (691 to 751 nm). As well as this spectrum of wavelengths, recent developments have created probes which change colour over time (fluorescent timers) and probes that can be 'switched on' to change their properties when irradiated with a certain wavelength (photoactivatable red fluorescent proteins—PARFPs). Other probes can be switched from red to green fluorescence (and back) at will (photoswitchable FPs). When incorporated into the advanced techniques of fluorescent microscopy these novel fluorescent proteins have been particularly useful in extending our understanding of cellular functions.

Identification of cellular components

The number of fluorescent dyes which bind directly to specific cell components is still relatively few compared to the tens of thousands of different proteins within a cell, a situation which has led to a technique called immunofluorescence. This uses molecules called antibodies, which are produced by our bodies as a form of chemical defence, which work by binding to proteins termed antigens on the surface of infective agents such as viruses or bacteria. Antibody–antigen binding is a highly specific reaction, often likened to a lock and key relationship, so that antibodies to chicken pox will not protect against measles. It is beyond the scope of this VSI to go into any depth in the production and use of antibodies (see *Viruses VSI* by D. Crawford), but given a target protein, or even part of a protein, an antibody can be relatively easily produced by injecting it into an animal such as rat, mouse, or rabbit—even sheep and goats—then biochemically isolating it from the blood. With direct immunofluorescence, the harvested antibodies are chemically conjugated (attached) to a fluorochrome, and the conjugate incubated with the tissue, before being thoroughly washed. If the antigen is present, the fluorescent antibody will bind to it, and produce a fluorescent signal when excited by the appropriate wavelength.

A crafty way round the time consuming production of fluorescent conjugates for every new antibody is to use indirect immunofluorescence, which makes use of the fact that serum antibodies of a particular species are constant, so that the antibody itself does not have to be labelled directly. In this technique the antibody which binds to the target protein (primary antibody) is not itself labelled, but a second fluorescently labelled antibody is used to bind to the primary antibody. Because the primary antibody has multiple sites to which the secondary antibody can bind, several secondary antibodies will attach, providing a brighter signal. For example, an antibody to a human protein that has been raised in mouse will recognize and bind to that protein in human tissue. This antibody could be directly labelled, but it is simpler (and easier) to use a labelled second antibody which reacts with all mouse antibodies (which has been raised in another species such as rabbit or goat). This saves the considerable effort of labelling each and every individual antibody, and secondary antibodies are usually commercially available with a choice of fluorochromes permitting multiple sites to be labelled in the same cell with different colours.

A gift from a jellyfish

In 1955, Osamu Shimomura, a research assistant at Nagoya University, was set a problem by his boss, Professor Hirata, which was to discover why the remains of a crushed mollusc, *Cypridina*, glowed when moistened with water. Shimomura successfully isolated the bioluminescent protein responsible, and, having published his results, was offered a position at Princeton University by Frank Johnson. Together with his new boss, they studied the jellyfish *Aequorea victoria*, which glows green around the edge when agitated. By 1962, they had isolated and published information on two proteins, the first of which they called aequorin, which had a mainly blue fluorescence; the second protein had a special chromophore which fluoresced green in UV light, and (unsurprisingly) came to be known as green fluorescent protein (or GFP). In 1988, Martin Chalfie, working on the development of *Caenorhabditis elegans*, a small roundworm which has 959 cells, realized that GFP would be the perfect tool to map the development of the worm's nerve systems. Chalfie got a researcher in his lab to sequence the protein, and then genetically manipulated this sequence into the protein responsible for the development of touch receptor neurons in *C. elegans*. When illuminated with UV, the cells containing the protein to which GFP had been added glowed bright green. This technique took off and, within a few years, GFP had been inserted into hundreds of proteins, perhaps the most spectacular being the fur of a pet rabbit called Alba, which glowed green in UV light. From this point, Roger Tsien used DNA technology to exchange various amino acids in different parts of the GFP molecule, producing a whole variety of fluorescent colours such as cyan, blue, and yellow. GFP and its derivatives became a standard tool for studying protein production and location throughout the scientific world, with green fluorescent organisms produced from humans cells to entire mice and pigs. In 2008, Shimomura, Chalfie, and Tsien were awarded the Nobel Prize for Chemistry. Today's fluorescent proteins, which also include those sourced from corals, make up a kaleidoscope of coloured probes, used throughout biology and medicine, but perhaps most notably in tracing neuronal pathways. Mice have been genetically modified to produce varying amounts of red, yellow, and cyan fluorescent proteins within the nerve cells of their brain, producing microscope preparations glowing with an amazing spectrum of colour, enabling individual cells to be traced, and leading to the whole system becoming known as 'the brainbow'.

Fluorescent probes from nanotechnology

Despite the sophistication of both the sources of excitation, their computer controlled rapid mechanical switching of filters and the spectrum of fluorescent probes, fluorescence microscopy fights a constant battle against fading, the progressive loss of signal over time as a result of a process called photo oxidation. Around the turn of the 21st century, a new class of fluorophore called quantum dots was introduced whose excitation and emission is fundamentally different, resulting in probes that are twenty times brighter and 100 times more stable than traditional fluorophores. Quantum dots are made from a variety of semiconductor materials, the most common of which are CdSE/ZnS, which have a core of cadmium selenide 10 to 50 atoms in diameter (about 100 to 100,000 atoms in total) with a thin film of zinc sulphide and an organic polymer which both makes them water soluble and enables bioconjugation with secondary antibodies. The colour of their fluorescent emission is determined by the size of the quantum dot, unlike traditional fluorophores, so that a single exciting wavelength can generate several colours in multiple labelling experiments. Quantum dots are also electron opaque, making them visible in electron microscopy, and consequently perfect for correlative light and electron microscope investigations. Although the presence of cadmium within the quantum dot core initially led to concerns over their toxicity, subsequent studies showed quantum dot toxicity to depend on a multiplicity of factors, including size and the use of various biological molecule encapsulating layers. Ideally, quantum dots in the future will be manufactured to order, either non toxic for cell biological investigations, or deliberately toxic for killing tumour cells which possess suitable binding sites for quantum dots to be accumulated preferentially while producing little or no effect on normal adjacent cells. Even without a toxic effect, it has been shown that if sufficient quantum dots can be attached to a tumour cell, excitation with a laser will cause sufficient heat build up in the quantum dots to kill the cell. Quantum dots are some way from a routine therapy in medicine and are one of many nanoparticles currently in development for targeted drug delivery, but their property of emitting strong fluorescence with the deep penetration of near infrared light can pinpoint tumours in the patient so as to allow image based surgery.

Advanced fluorescence microscopy

As well as generating a whole new group of acronyms, the development of fluorescent probes has neatly complemented the advances in the imaging of their signals to produce what can only be described as a rebirth of light microscopy, particularly in cell biology. Whilst the whole of the rest of this book could be spent describing these novel technologies, we will attempt to briefly review the progress and possibilities in this area of fluorescent microscopy.

As we have seen from Chapter 3, the irresistible progress in microscopy has breached the diffraction barrier identified by Abbe in the 1880s, extending the range of optical microscopy to the identification of a single molecule. Although the first account of a technique called TIRF microscopy was published by E. A. Ambrose in 1956, it was not incorporated into fluorescence microscopy until the early 1980s by Daniel Axelrod. To detect emission from a single fluorophore, the background (and also 'noise' from the detector) must be extremely low. This can be achieved by illumination from the side, with the specimen surrounded by materials of high refractive indices. Most of the illuminating light will bypass the objective lens, but a small proportion which hits the specimen at a suitable angle will be totally internally reflected, creating what is termed an evanescent wave of high intensity against a low background. This light only penetrates a very short distance into the specimen (200 nm), making the system ideal for studying membrane events at the surface of living cells, and allowing the movement of single molecules across the membrane surface to be followed for the first time. This method of imaging can also be combined with the novel technologies of Fluorescence recovery after photobleaching (FRAP) and Förster resonance energy transfer (FRET), producing a very powerful tool for the study of the behaviour of individual molecules.

If an intense spot of exciting light is applied to fluorescing molecules, they will very quickly become bleached, losing their fluorescent signal. When just part of a cell is bleached, the fate of adjacent unbleached molecules can be followed by simply watching what happens next. If the bleached area stays black, then those molecules are not mobile. If however they are moving, either by active processes or diffusion, they will move into the bleached area, and restore the fluorescence. This technique is known as fluorescent recovery after photobleaching. By measuring the rate and amount of fluorescence recovery (usually plotted as a graph), the rates and speeds of individual molecules as they travel through the cell can be established. A similar approach (albeit by a different mechanism) is to use a photo-activated probe, which is allowed to permeate the whole cell. By 'switching on' the fluorescence with a spot of light in a small area, and then watching the spread of fluorescence from that point, transport rates can be similarly determined, a system called FPA (fluorescent protein activation).

Despite its numerous advantages, the signal generated by the fluorescence microscope is essentially a brightly coloured dot on a black background, meaning that there is no local context for the signal. As we shall see, this difficulty is overcome at the molecular level by electron microscopy, but not in living cells. Fluorescence can, however, tell us about the immediate neighbours of the fluorescing molecule, as a result of a property described by Theodor Förster. Förster showed that if there is a second fluorophore (of a different colour, let's say green) close enough to the primarily excited fluorophore (let's say red), when the energy is lost from this molecule, it can cause the neighbouring molecule to fluoresce and produce a green signal. The neighbouring molecule (the acceptor) has to be within 10 nanometres of the original molecule (the donor) for this transfer to occur, and indicates that the molecules were interacting. This approach can also be used as a negative indicator, where molecules that might have been interacting can be tagged with red/green fluorophores, and shown not to interact in the absence of the switching of red to green fluorescence. Förster resonance energy transfer (FRET) has consequently become a powerful tool in cell biology and also a basis for biosensor technology.

Whilst a molecule is actually emitting fluorescence, it will be in an excited state. Once the photon of fluorescence has been released, then the molecule will relax down to the ground state. Using a pulsed source of excitation, the fluorophore will be repeatedly 'switched on'. The time for which the molecule is actually excited is called the fluorescence lifetime. This is quite brief—around 5 to 10 nanoseconds (a half to one hundred thousandth of a second). In an isolated molecule this lifetime is determined purely by the properties of the molecule, but when any molecule is surrounded by other molecules (as it always is in a cell), then dynamic interactions with neighbours (see FRET) can shorten this time, and the excited energy may also be lost as heat (fluorescence quenching), which also shortens the fluorescence lifetime (returning quicker to the ground state by emission of another photon). A technique called time correlated single photon counting is used to generate curves of fluorescence lifetime to give information of the environmental effects on fluorescence of the molecule in question. A combination of FLIM (fluorescence lifetime imaging microscopy) and FRET is used to make biosensors which monitor metabolic activities within cells. Perhaps the simplest example is the use of a biosensor attached to an enzyme that cleaves proteins, separating them to a point that they are no longer close enough for fluorescence energy to occur, and consequently causing a reduction in FRET.

Multiphoton microscopy

Pulsed lasers produce incredibly short bursts of electromagnetic energy. For example, a pulsed femtosecond laser produces a flash of light that lasts for femtoseconds to a picosecond (a picosecond is one trillionth of a second, a femtosecond is one thousandth of a picosecond), instantly followed by another (and so on). These lasers brought about the possibility of exciting fluorophores with two photons of only half the necessary energy, but they need to arrive almost simultaneously to generate the ejection of a photon. Infrared pulsed lasers penetrate living tissue more effectively, with the advantage that fluorescence is achieved from much deeper in the tissue than normal fluorescence, where the depth of penetration is limited by multiple light scattering events. Multiphoton microscopy (mainly two photon in practice, but also feasible as three or more photons) allows imaging from as deep as a millimetre (one thousand micrometres), an improvement of several hundred micrometres over fluorescence confocal microscopy. A second advantage of two photon excitation is that it forms as a single spot in the axial plane (z axis) without the 'hourglass' spread of out of focus light (the point spread function) that happens with single photon excitation. This is because the actual two photon excitation will only occur at the highest concentration of photons, which is limited to the focal plane itself. Because there is no out of focus light, there is no need for a confocal pinhole, allowing more signal to reach the detector. Combined with the increased depth of penetration, and reduced light induced damage (phototoxicity) to living tissue, two photon microscopy has added a new dimension to the imaging of living tissue in whole animals. At the surface of a living brain, remarkable images of the paths of whole neurons over several hundred micrometres can be reconstructed as a 3D z section from an image stack imaged through a thinned area of the skull in an experimental animal. Endoscopes have been developed which incorporate a miniaturized two photon microscope, allowing deep imaging of intestinal epithelium, with potential to provide new information on intestinal diseases, as most of the cellular lining throughout our gut is thin enough to be imaged in this way. So far a whole range of conditions including virtually all the cancers of the digestive tract as well as inflammatory bowel disease have been investigated, reducing the need for biopsies and providing new insights as to the nature of these conditions.

The combination of these emerging techniques applied to cell biology and medicine is producing a similar flood of new discoveries to those in the time of Pasteur and Koch, only this time at a molecular level. Combined with the promise of nanotechnology in general and in medicine in particular, there should be significant advances in therapy of many diseases in the relatively near future.

Chapter 5
Electron microscopy and the dawn of atomic resolution

On Christmas day 1906, Ernst Ruska entered the world, the son of a professor in Heidelberg, a centre of German and international academic excellence both then and now. Ruska's father had a large Zeiss microscope, a source of fascination for Ruska from a young age, as were the astronomical telescopes at the nearby observatory, run by his uncle. Ruska subsequently studied electrical engineering, and became a member of a team led by Max Knoll, helping to develop the cathode ray oscilloscope, in which a heated metal source generates a wide beam of electrons which are focused by two short electrically driven magnetic coils to form a small spot on a fluorescent screen. In the 1920s, Hans Busch had suggested that these magnetic coils were acting on the electron beam in the same way that glass lenses focus light beams. At the time, it was not known exactly how electrons behaved, as they were thought to be a stream of tiny particles, unlike the wave nature of light, but in 1924 Louis de Broglie, a French physicist, devised a theory suggesting that minute particles (electrons) could behave as waves, and consequently be diffracted. Ruska was aware of these findings, but as the wavelength of electrons was unknown at the time, he worried that there might be little or no advantage to using electrons for microscopy. Applying de Broglie's equation, however, showed that the wavelength of electrons was ten times smaller than light (actually a thousand times smaller), meaning that electron microscopy should provide an improvement of one thousand times more than the resolution of light microscopy. Ruska went on to pioneer electron lens design, and in 1933 he built an electron microscope with a magnification of 12,000 times. Lacking the necessary funding to move his work to the next level, he moved from academia to the German electronics giant Siemens, and in 1939 the first commercially produced electron microscope was installed. Ruska often stood alone in the face of a sceptical scientific community with respect to electron microscopy, but was entirely justified, receiving the Nobel Prize for Physics in 1986 at the age of 80. In his acceptance speech he commented that 'the doubt of others had the advantage of leaving the field uncrowded'. Ruska had survived long enough to enjoy his achievement, if only briefly, and died in 1988.

How to make a transmission electron microscope

Electromagnetic lenses focus electrons as glass lenses focus light. Consequently, all that is required to produce a transmission electron microscope is to fulfil the conditions necessary to transmit a beam of electrons through a specimen and produce a magnified image. The first prerequisite is a beam of electrons, which are generated by heating a thin V shaped tungsten wire to around 1,500 degrees Centigrade. In air, electrons are instantly absorbed by the local molecules of the atmosphere, so that a high vacuum environment is required to permit the electrons to travel down the column, pass through the specimen, and be formed into a magnified image. Electrons radiate in all directions and need to be formed into a directed beam by creating a potential difference between the tungsten wire filament in the cathode, at the top of the microscope column, and an anode plate, positioned below. This potential difference (the accelerating voltage) is usually in the order of 50,000 volts. The microscope column is around two metres in length, orientated 'upside down' with respect to a light microscope, as the illuminating electron source is at the top, which makes alignment of the beam more convenient (Figure 13). Electromagnetic condenser lenses focus the electrons onto the

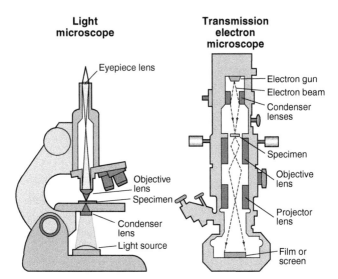

Light microscope

Eyepiece lens

Objective lens

Specimen

Condenser lens

Light source

Transmission electron microscope

Electron gun

Electron beam

Condenser lenses

Specimen

Objective lens

Projector lens

Film or screen

13. Comparison of the light and electron beam pathways in conventional light and transmission electron microscopes.

specimen (mounted in the centre of the column), and a magnified electron image is formed by objective and projector lenses below the specimen. Obviously, we cannot see electrons, but a beam of electrons (like light) can generate fluorescence, and react with a phosphor coated metal plate to provide an image, in which contrast comes from the different absorption and interactions of electrons with different parts of the specimen. The screen image is viewed through a glass window at the base of the column, often through a x 10 binocular microscope. Most modern microscopes incorporate a camera sensor onto which the beam can be focused, and the image can be viewed on a monitor. For the first few decades of transmission electron microscopy, however, operators would sit in a darkened room, peering down the binoculars, altering the magnification, and focusing the beam by changing the current in the lenses, then lifting the screen to allow the electron image to expose photographic emulsion (again just like light) and record the image. The specimen is moved around manually by a pair of micrometre screw gauges driven by rods at the base of the column, precisely engineered to allow tiny movements of the specimen at magnifications in excess of one hundred thousand times. Modern instruments generally have motor driven stages, allowing both xy movement (side to side and up and down) and tilting of the specimen to permit 3D reconstruction (EM tomography).

Although the last paragraph points out the relative simplicity of making an electron microscope, the day-to-day running could often test the patience of even the most seasoned operator. The need for a high vacuum environment requires continuous evacuation of the column to one millionth of the surrounding atmospheric pressure, which requires an involved pumping system creating increasing levels of vacuum. In order to avoid letting the entire column down to air at every specimen change (and waiting for an hour to pump out again), the specimen chamber is isolated by means of an airlock, requiring only two to three minutes to change the specimen. Other bugbears of early instruments were water leaks from the cooling systems in the lenses, which consume large amounts of power and need to be cooled to a constant temperature for stability. The generation of the high accelerating voltage that was constant was not without problems, and variations in the electron optics required constant adjustments to produce an image free from distortion and astigmatism. However, all these aggravating factors were more than compensated for by the satisfaction of producing a micrograph that was sharp and well contrasted, at the end of a total process from the original material that could take several days from start to finish.

EM specimen preparation

A high vacuum environment and the intense irradiation of an electron beam preclude observation of living material by transmission electron microscopy. Some organisms might (briefly) survive high vacuum, but even if this was not a problem the radiation damage generated by the electron beam would be enough to quickly kill most living material. Because the electron beam has limited penetration, the specimen has to be extremely thin for the beam to pass through. Material science specimens such as metals or rocks are polished down to a sliver and biological material needs to be specially processed to produce ultrathin sections. For light microscopy, biological material is chemically stabilized (termed fixation) by immersion in formalin (an aqueous solution of formaldehyde), followed by removal of water by immersion in successively higher concentrations of alcohol (ethanol) and infiltration with wax to support the tissue, allowing sections a few microns thick to be cut on a microtome. Sections are mounted on a glass slide, stained to produce contrast, and viewed in the light microscope. In surgical situations (while the patient is still on the operating table), a biopsy can be instantly frozen and sectioned from the frozen state so that the pathologist can make a rapid diagnosis to instruct the surgeon.

Before sectioning techniques for EM emerged in the 1950s, a popular way for electron microscopists to look at surfaces of bulk objects was by carbon replicas. The surface of interest was coated with a very thin layer of evaporated carbon under vacuum (by passing current through carbon electrodes) and then a metal such as platinum was evaporated onto the replica from a low angle, so that it settled on the surface of the carbon rather like a snow drift, producing a 'shadowed' effect to produce contrast once the replica had been separated from the surface and examined in the microscope. For biological material, specimens were frozen, split open and etched by the vacuum (removing surface water), then replicated in a technique known as freeze etching. Strong alkali (usually in the form of domestic bleach) was used to clean off the original material, and the replica viewed with the transmission EM.

Sectioning for EM was developed as a modification of LM techniques. Small blocks of tissue around a one millimetre cube are fixed in glutaraldehyde (related to formaldehyde), a chemical which cross-links proteins to stabilize them. Because the electron beam can only produce black and white images, contrast is produced by heavy metal solutions, both as fixatives and stains. Osmium tetroxide, a highly effective oxydizing agent, is used as a second fixative after glutaraldehyde. Osmium has a particular affinity for fats (lipids) which are the main component of membranes, which appear in micrographs as thin black lines which surround every cell. In the late 1940s and early 1950s, pioneers such as Keith Porter and George Palade at the Rockefeller Institute, and Fritiof Sjöstrand in Sweden, found solutions which are still used to this day. They modified microtomes to cut sections one hundredth of the thickness of standard wax sections, and infiltrated with resins such as araldite to support the tissue during ultrathin sectioning and withstand the effects of the electron beam. Metal microtome knives were replaced with carefully broken edges of plate glass or specially sharpened diamond. They cut 'ultrathin' sections, which produce interference colours from ambient light. Silvery grey sections indicated an optimal thickness of 40–60 nanometres, with gold interference around 100–20 nm thick. A small water bath is attached to the knife so that the freshly cut sections float on the surface, at which point they are picked up on a small copper grid 3 mm in diameter, and then the grid (with attached sections) is stained in solutions of heavy metal salts (uranium acetate and lead citrate), dried, and transferred to the specimen stage for viewing (Figure 14).

Porter and colleagues were rewarded for their efforts by a quantum leap in the visualization of cell structure. The indistinct and irregular shadowy lumps of light microscopy became clearly defined bodies with internal structure displayed at a level that was previously unimaginable. In the words of another pioneer, Don Fawcett, 'for morphologists, the decade from 1950 to 1960 held

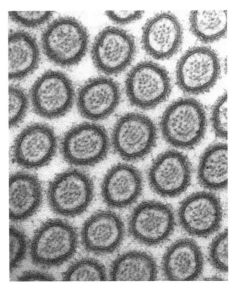

14. A typical TEM thin sectioned image, showing a section through the tiny fingerlike projections at the surface of our intestines (microvilli) which take up the nutrients from ingested food. Each profile has a double membrane, and supporting protein filaments in the centre, measuring 50 nanometres in diameter, here at a final magnification of around half a million.

the same anticipation and excitement that attends the opening of a new continent for exploration'. Fawcett's Atlases of Ultrastructure are classics, and still bear suitable comparison to work produced fifty years later. Not only was electron microscopy effective in showing new vistas of structure, it was also aesthetically attractive in the extreme. The symmetry and ordered level of structures within our cells and tissues and their consistency throughout all living creatures was (and still is) a marvel to behold. Palade, together with Albert Claude and Christian de Duve, was awarded the 1974 Nobel Prize for Physiology or Medicine 'for their discoveries concerning the structural and functional organisation of the cell'.

How true to life are the structures shown by electron microscopy?

Despite the avalanche of information produced by the paradigm shift of electron microscopy, questions surfaced as to how representative a chemically stabilized, dehydrated, resin infiltrated, ultrathin sliver of a once living cell actually was. As water is a major cell constituent, its removal produces volume shrinkage of up to 50 per cent. Is there any way that electron microscopy of living materials can be confirmed by a more direct approach? Fortunately there is, and in one form or another it has been about since the earliest days of electron microscopy, and is dependent upon the aqueous nature of our cells, which can be stabilized and rigidified in one step, by freezing. However, if water is frozen slowly, individual ice crystals will nucleate and grow to the extent that they distort cell structure badly, but if the water in cells is frozen almost instantly, ice crystals do not have time to form, and the ice formed is amorphous, with individual molecules randomly arranged, as they are in glass, a process called vitrefication. In the earlier days of electron microscopy, frozen sectioning techniques had not been developed, but the frozen material could be fractured with a razor blade, some of the ice sublimed (removed directly as vapour), and a thin replica made by evaporating carbon under vacuum onto the fractured surface. The replica was then 'shadowed' by evaporating metal particles on it at a low angle, floated off the specimen, and examined directly in the transmission microscope. This technique is called freeze etching, and can produce detail at the molecular level of components, remaining to this day as a powerful technique as developed by John Heuser as quick-freeze deep-etch.

Cryo-electron microscopy

Fast freezing of material without chemical intervention has become a more mainstream part of biological electron microscopy in the last decade, after a lifetime of persistence by pioneers such as Jacques Dubochet. Rapid freezing is achieved by plunging specimens into liquid ethane cooled in liquid nitrogen (at –196 degrees C), or by directly subjecting material to high pressure and a spray of liquid nitrogen in a special apparatus. Using special microscope specimen stages cooled by liquid nitrogen or liquid helium, frozen sections from the cryo-ultrotome can be viewed at very low temperatures, which protects them from beam damage. This is a technique termed CEMOVIS (cryo-electron microscopy of frozen hydrated vitreous sections). Alternatively a thin film of an aqueous suspension of purified single molecules can be rapidly frozen and examined, and despite their random orientation in the frozen film, multiple images can be collected and averaged in the computer, producing a combined result of near atomic resolution. It is remarkable (and reassuring) that throughout the development of specimen preparation techniques over the entire history of biological electron microscopy, whether chemical or freezing approaches (or a mixture of both) have been used, there have been broadly consistent findings. Advances are continuously made in extra resolution and detail, but there has been virtually nothing of significance in cell structure that has been shown to be a result of preservation technique (i.e. an artefact) rather than an actual structure.

High voltage electron microscopy

The depth to which an electron can penetrate a specimen is determined by its accelerating voltage. In the 1960s, manufacturers in Japan and the UK responded to the demand for increased penetration by producing instruments with a million volt accelerating voltage. Moving to this level was impressive, as the generation of one million volts is in itself certainly not routine technology, requiring an entire building to be purpose built to house the instrument with the high tension generators occupying the first floor and the microscope sitting beneath them (Figure 15). The increased beam energy requires more powerful lenses,

15. A million volt microscope and cut away diagram. Note the scientist at the level of the column for an idea of scale.

increasing the diameter of the column from 25 centimetres to in excess of one metre, with the height of the column approaching 4 metres. This was reflected in the cost, as conventional instruments at the time were around £25,000, whereas a million volt instrument cost £1 per volt. High voltages generate considerably more radiation, requiring the leaded glass for the viewing window at the base of the column to be increased to a thickness of 20 cm instead of one centimetre in conventional instruments. Radiation as a hazard for the EM operator is not a major concern with conventional instruments, but it is standard for users to wear a radiation dosage monitor even when using a million volt machine for a short period (although the manufacturers supply adequate shielding). Million volt microscopy has the advantage of a considerable increase in specimen penetration, allowing biologists to increase the depth of their sections to around one micron, giving a slice that equates to around 10 per cent of the volume of an entire cell, compared to the usual one thousandth. Another advantage is that because less energy is deposited in the specimen by the higher accelerating voltage, there is actually less beam damage than usual. However, there is ten times the content in the thicker sections, superimposing structure upon structure in a way that is difficult to interpret, although more recent instruments use a tilting specimen stage to produce 3D images which permit depth perception. It is fair to say that ultra high voltage results for biological material have been interesting rather than revolutionary, but in materials science and nanotechnology, the increased penetration power has produced more remarkable information. The cost of installation and maintenance of these (now relatively rare) instruments tends to make them national rather than institutional facilities. Funding bodies rightly insist that the costs of these facilities require that they be made available to scientists from far and wide. Many of the original advantages of million volt instruments have now been incorporated into a newer generation of microscopes which operate at lower beam energies (up to around 400 kilovolts).

The physics of beam/specimen interactions

So far we have concentrated on the basic similarities of electron and light microscopes, and the conventional form of imaging of thin sections in a transmission instrument, in which a contrast in the image is produced by the interactions between the beam and different parts of the specimen. The physics of electron/specimen interactions shows them to be diverse, producing a variety of characteristic (and useful) signals which can be collected in order to generate further information about the nature of the specimen. When electrons travel through the specimen, they either pass straight through completely unaffected (in which case no information or image is generated), or they are scattered, and emerge from the exit surface of the specimen in a non-uniform distribution, containing both structural and chemical information. Scattering happens as two alternatives, either elastic events or inelastic events. An elastic event changes the trajectory of the electron, but does not affect its velocity or kinetic energy. If electrons pass through a regular crystalline array of atoms or molecules in the specimen, the elastic scattering will produce diffraction patterns (like the dots we see when looking through the fine fabric of an umbrella at a street lamp). These diffracted electrons can be focused into a pattern of dots and their distribution gives information about the arrangement of atoms within the specimen at high resolution. It was the electron diffraction images of DNA produced by Rosalind Franklin that were the key to Watson and Crick's discovery of the 'structure of life'. Another major consequence of elastic events is the production of backscattered electrons, which we will consider when we describe scanning electron microscopy.

Inelastic events are those in which there is a transfer of energy from the electron to the specimen, which leads to the generation of secondary electrons, Auger electrons, X-rays, and even photons of visible light (see Chapter 5). As with backscattered electrons most of these interactions are more important in scanning electron microscopy, but the production of X-rays, along with electron energy loss, provides signals that can be used to map the chemical composition of the specimen in the transmission EM.

Analytical electron microscopy

When the electron beam hits a particular element in an inelastic collision the interaction produces two types of 'fingerprint' specific for that element. One is the production of X-rays, and the other is the energy loss suffered by electrons. For X-ray detection, an energy dispersive X-ray spectrometer (EDXS) is fitted to the column, with a detector in the specimen chamber. For measurement of electron energy loss (EELS), two electron spectrometers are mounted on the column. Not every analytical microscope will be fitted with both systems, as they will both produce information about the amount and distribution of individual chemical elements in the specimen, but they have different attributes, in that EDXS (X-ray detection) tends to be better for analysis of heavier elements, and is easier to use, whereas EELS is better for elements with low atomic numbers, from carbon up to zinc. EELS is also more sensitive, and can detect different forms of the same element. Using EDXS to analyse the distribution of metals in an alloy, a thin sample is imaged in the microscope, and then the beam is focused onto individual areas, and signals are collected which show both the composition and amounts of the elements in that area. The beam is then focused on a different area, and, if required, a whole area can be mapped. Alternatively, in a microscope which is constructed specifically for energy filtered imaging (EFTEM), a magnetic prism spectrometer and an energy selecting slit are incorporated into the base of the column, producing an image via a cooled CCD camera in which the selected element shows up as a bright area against a dark background. In materials science, the elemental make-up and distribution in the specimen is clearly crucial to the performance of the alloy or the components of microelectronic circuitry. In life sciences, as most biological material is predominantly made of four elements (carbon, hydrogen, oxygen, and nitrogen), analytical electron microscopy is more specialized, but can still be extremely useful. An example from as early as 1982 demonstrated the accumulation of copper in the cells of the liver, brain, and cornea of patients suffering Wilson's disease. Also, because virtually all the phosphorus in a cell is in DNA, if an image in which the phosphorus content is compared with that of nitrogen (spread evenly throughout the cell) then the distribution of DNA in the nucleus can be accurately mapped. This approach is known as energy filtered transmission electron microscopy (EFTEM).

From two to three dimensions—electron tomography

It has been estimated that a career electron microscopist who spends his working days preparing, sectioning and staining, observing, and recording biological material will get through the equivalent of one cubic millimetre of tissue in a forty-year working lifetime. Although every thin section is packed with information, it is still a fragmentary sample of the whole tissue. Naturally the investigator will look at many different cells in the section, to convince himself that any findings are truly representative of the whole cell population. One way to extract extra information is to record and compare successive sections through the tissue, and attempt to build up a three dimensional representation, originally by tracing outlines on perspex sheets and then piling them on top of each other. Unlike wax sections in light microscopy, which form a ribbon of successive slices through a cell, making serial sections from a resin block is tricky, making attempts to reconstruct a three dimensional view of cell contents extremely difficult. Around the turn of the 21st century, however, the development of tomography in the transmission EM has enabled 3D reconstruction of cell components at molecular resolution. TEM tomography works on the same principle as the more familiar X-ray tomography where X-rays are taken at different angles through the patient, and the recorded images processed by computer to produce a three dimensional reconstruction which can be viewed in various ways (usually as a series of slices). Imaging by X-rays keeps the patient still and moves the imaging mechanisms. In the electron microscope, a thick section (more content) viewed at high accelerating voltage (more penetration) is tilted between angles of plus 70 degrees to minus 70 degrees with a mechanically driven, computer controlled specimen stage, and an image is recorded after each tilt by a camera. Once a suitable area of the specimen has been selected, the operator sets the imaging collection software in motion, and leaves the hardware for the next few hours or so to record an image, gently tilt the specimen, record the next image, and repeat until a data set with 140 images has been collected. It has been the refinement in both the accuracy of tilting specimen stages (some of which tilt in more than one direction) and the improvement in digital cameras, together with continuously improving software, that now enables three dimensional reconstructions of parts of the cell at a level of detail probably unimagined by the pioneers of electron microscopy. That electron microscopy is still an emerging discipline is demonstrated by the direct link with tomography expert Richard McIntosh, who was a research student with Keith Porter in 1968, and subsequently succeeded him as lab head at Boulder, Colorado, where Porter had installed one of the first million volt microscopes. McIntosh himself retired as director in 2007, but like most many microscopists, still remains an active scientist. In a turn of the century article in the prestigious *Journal of Cell Biology* (founded by Porter and others in 1955), McIntosh pointed out that three dimensional EM (Figure 16) will provide an important complement to the emerging light microscopy of living cells, as the spatial resolution (at the time) was fortyfold better than confocal light microscopy. To quote: 'We see a brave new world emerging from the combination of modern EM with the rapidly advancing methods for light microscopy.'

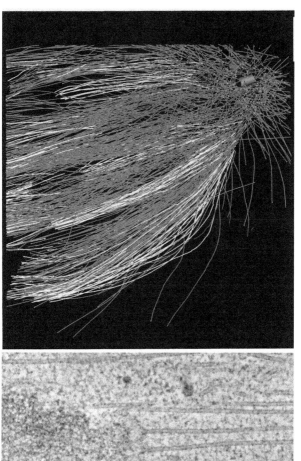

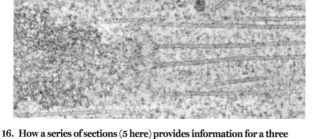

16. How a series of sections (5 here) provides information for a three dimensional reconstruction. The lower image shows a few pipe-like profiles of microtubules, part of the apparatus for separating chromosomes in division, and the upper image is a reconstruction of the entire area, with the position of numerous microtubules indicated.

Labelling in transmission electron microscopy

As we have seen in Chapter 4, the fluorescence microscope, in various advanced forms, has broken the 'diffraction barrier', and can now produce detail with a resolution in living cells that is now much closer to electron microscopy. At this point the obvious question is 'Why bother with electron microscopy at all?' The simple answer is to use the (miniaturized) black cat in the cellar analogy. With fluorescence microscopy, a light beam would pick out the reflection from the cat's eyes but little else, whereas with electron microscopy the entire cat would be visible with each strand of fur resolved (although our mythical miniature cat would have to have been suitably preserved). The bottom line here is that with electron microscopy, all the specimen is visible at molecular resolution in the same context, whereas with fluorescence, the image is formed from a collection of coloured (fluorescent) dots against a black background. Although the imaging mode in the fluorescence microscope can be flicked back at the press of a button to phase or interference contrast for a general view of the cell, this also means a return to conventional levels of light microscopy resolution. At no stage should we get into an EM good/LM bad (or vice versa) argument, which is completely pointless, but we should view them as complementary techniques which can be combined to produce a synergy of discovery. Electron microscopy (despite its extensive preparation) shows exactly where a labelled molecule is situated, together with the surrounding molecular organization *in situ*. To get the best of both worlds, we will consider correlative microscopy (of LM and EM) in the section 'Merging LM and EM'.

In the absence of a bright fluorescent spot of colour, however, how do we recognize a molecular tag in the electron microscope? Also, in a monochromatic electron imaging situation, how can we discriminate between different labels? Given the black and white nature of an electron microscope image, we require a label (targeted to a specific area by antibodies in the same way as LM) which will stand out with respect to two parameters, namely contrast and shape. Intense contrast is achieved by using a marker which is electron dense, and for many years tiny particles of colloidal gold have been coated with an antibody, to stand out as an unequivocal black dot on a grey scale image. Making a suspension of colloidal gold particles by reducing chloroauric acid was, for the early electron immunologists, a little like the conversion of base metals to gold by the medieval alchemists. Gold is already present in the chloroauric acid in the form of ions, and chemical reduction converts it to regular molecular aggregations of metallic gold, creating a suspension of tiny gold particles which disperse light, producing the deep red colour of a fine claret in a successful reaction. The size of these aggregations is dependent on the rate of reduction, enabling colloidal particles to be produced at sizes of one to forty nanometres, offering a method to distinguish different sites of labelling by the size of the gold colloid. Gold labelling in electron microscopy could be considered a bit of a black art, but always produced disproportionate satisfaction when black dots were seen indicating a specific molecule—probably due to the non cognoscenti gibe of 'pretty pictures' often being the endpoint of much effort by the microscopist. After many years of the round profiles of colloidal gold tags, colloids of other metals have been introduced, which have a similar size (and similar labelling activity), but different shapes, such as the faceted colloids of palladium, which can have faces which are sharply geometrical, such as triangles, hexagons, or pentagons, or can be umbonate in shape (like popcorn!). We should also include quantum dots (see Chapter 4) from nanotechnology, engineered on the nanoscale to produce fluorescence in the light microscope, but with sufficient metal from the cadmium content to be electron dense in the electron microscope. As with other colloidal electron dense markers, different sizes can be used for double labelling. One further 'close to magical' property (as far as the working microscopist is concerned) is that quantum dots, unlike the vast majority of fluorophores, maintain their ability to fluoresce throughout the processes of fixation, dehydration, resin embedding, and thin sectioning, enabling selection at the light microscope level of the best labelled areas for electron microscopy.

Merging LM and EM—correlative microscopy

Using LM and EM together is clearly a beneficial approach, for example tracking the dynamics of internal membrane systems in light microscopy, but needing EM resolution of these events, which can only be visualized after stabilization by fixation or freezing as a start to the somewhat involved EM preparation protocol. As most life science microscopists use both EM and LM routinely, putting them together was not a revolutionary thought. The main barrier to looking at the same cell in LM and EM is one of relocation—in the early days the equivalent of finding the same needle in a haystack twice. However, Ralph Albrecht in the 1980s and Alexander Mironov in the 1990s showed that it was possible. Albrecht combined light microscopy of platelets (which are cells involved in blood clotting) with both high voltage transmission EM (avoiding sections as the beam would pass through the whole cell) and scanning EM to visualize the distribution of surface receptors and shape changes involved in the formation of a blood clot. He did this by mounting his living cells on a thin film of formvar (a tough electron transparent film), on the surface of a transmission EM grid (which was marked with a series of coordinates to locate individual cell positions). Having observed platelets changing shape in the LM, the grids bearing them (having also been labelled with colloidal gold) were plunged quickly into fixative, and prepared for transmission EM or scanning EM, producing a combined (correlative) series of results for cell behaviour (LM), cell shape (LM, TEM, SEM), internal organization (TEM), and surface morphology (SEM). Mironov went one step further in managing to fix and embed the very cell he was studying in time lapse LM for membrane dynamics, producing sections for TEM cut in the same plane as appeared in the light microscope, including fluorescent labelling with green fluorescent protein. This certainly comes close to an infinite capacity for taking pains (as is much of electron microscopy), but once the system is set up, the rewards are well worth the initial effort.

Since then, correlative microscopy has become relatively routine. Life has been made much easier by probes that retain their fluorescence through EM preparative protocols, so that green fluorescent labelling can be observed in the fluorescent microscope on a section that is directly transferred (via special localization holders) to the transmission EM, where the same area is located automatically with the software driven mechanical stage, and the green fluorescent probe LM image can be directly superimposed on the transmission image of the same cell in a matter of minutes. In some cases the optical systems of both light and electron microscopy have been combined in a single instrument. This combination of LM and EM allows investigations which range from whole tissue to individual molecules.

Chapter 6
The electron microscopy of surfaces

The importance of surfaces

Long before the idea of scanning specimens with a small spot of light produced confocal light microscopy, the idea of using a small spot of electrons to scan surfaces had been around for as long as electron microscopy itself. A surface demarcates the boundary of a solid, and is the site of interaction with the surrounding environment, from a ball bearing to a living cell. In the mechanical world, adhesion, friction, wear, and corrosion are all dependent upon surface properties. The smooth surface of a ball bearing is crucial in the reduction of friction, but its efficiency may well be compromised by wear or corrosion. Surface properties of lenses are routinely modified with coatings which alter optical properties. Surface coating may make surfaces hydrophilic (and water soluble) or hydrophobic (to repel water). In blood, white cells which are responsible for our immune defences rely on surface molecules to both recognize and respond to infective agents such as bacteria and viruses. In multicellular tissues, information is passed between cells via their surfaces, and nerve impulses also travel across the surface of neurones. Surface chemistry and physics encompass interfaces between solids, liquids, and gases, such as the catalysis of gases in car exhausts. In biology, catalytic reactions are brought about by enzymes acting at the surface of membranes. All adhesion relies on surface properties, be it interactions between a tyre and the road surface (enough friction to move the car or enough adhesion to prevent skidding?). Around 70 per cent of joints in aircraft rely on surface bonding, and we trust them to fly without falling apart. Although microscopy in general and scanning electron microscopy in particular are by no means the only ways of characterizing surfaces, they are a cornerstone of understanding their structure and properties.

Rather than passing a 'flood' beam of electrons through a thin sliver of material as in transmission EM, scanning microscopy moves a small spot of electrons to and fro in a pattern of horizontal lines over the surface of the specimen (a raster). At this point it is important to point out that resolution in the scanning microscope is not determined by the wavelength of electrons as it is in transmission instruments. The resolution is determined by the diameter of the spot of electrons, as it is this area (together with the depth to which the beam penetrates) which makes the 'interaction volume' from which the signal is produced. As the beam diameter is reduced, so is the interaction volume, and resolution is increased. This constraint caused scanning EM to be considered 'low resolution' by the purists of transmission EM, and it took around fifty years for scanning EM to come close to the resolution of transmission EM, although the most modern instruments are almost equivalent, despite the different mode of imaging (Figure 17).

How to make a scanning electron microscope

From the top of the column to the specimen, there is little difference between a scanning EM and a transmission EM. Both have a source of electrons at the top of an evacuated column, and a set of condenser lenses to focus the electron beam onto the specimen. The first generation of scanning microscopes used the same source of electrons as transmission instruments, a 'V' shaped tungsten wire heated to around 1,500 degrees C. Heating provides the energy required to remove electrons from the atoms in the

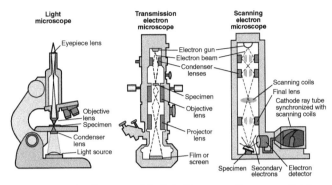

17. Comparative diagrams of the light/electron pathways on a light microscope, transmission EM, and scanning EM. The scanning and transmission instruments are similar until the electron beam hits the specimen.

tungsten, overcoming a barrier called the work function. If more electrons can be extracted, then a brighter and smaller spot can be used to scan the specimen, increasing the resolution. The first improvement in electron sources was to use a crystal of lanthanum hexaboride (Lab 6), which has a lower work function than tungsten, increasing the brightness by a factor of ten. The next stage, which produces a further hundredfold improvement in brightness over Lab 6, is to use a field emission gun (FEG). In field emission, electrons are extracted from an ultrafine needlepoint of tungsten by the application of a powerful electric field close to the tip which overcomes the work function, creating a process called electron tunnelling. This requires a vacuum of two orders of magnitude greater than conventional electron sources, requiring ion pumps in the electron gun. Field emission guns (FEGs) can be either 'cold' (at room temperature) or heated. Cold FEGs are the most effective, but can suffer from instability as a single molecule can contaminate the tip, and reduce emission. A practical compromise is to use thermal (or Schottky) field emission, in which the filament tip is both heated (which repels any potential contamination) as well as being subjected to an intense field. Most top of the range scanning (and many transmission) microscopes have field emission sources, and are termed FEG-SEMs or FEG-TEMs.

Having generated an intense beam of electrons, the next stage in scanning EM is the point at which everything changes from conventional transmission EM. Rather than illuminating the whole of the specimen in a 'flood beam', the electrons in the scanning EM are concentrated into a small spot (demagnified) by two electromagnetic condenser lenses, and focused onto the surface of the specimen by the final lens directly above the specimen area, sometimes (inappropriately) called the objective lens. The spot of electrons is moved across the specimen in a raster by a set of scanning coils which are usually contained within the final lens. These are four radially orientated electro magnets arranged so that their magnetic fields are perpendicular to the axis of the beam. By varying the current, the beam can be controlled to scan across (raster) the chosen area of the specimen, building an image dot by dot and line by line. Magnification in the scanning EM is determined by the chosen area of the scan as the final image is produced on a fixed size monitor. Thus if the scanned area is reduced by half, then the magnification will double. Most instruments will allow modest magnifications at the bottom end of their range (x 20 or so) up to 200,000 times in research grade microscopes. Because the image is built up from signals generated by the focused spot of the scanned beam, it is the diameter of this spot that ultimately limits the resolution in scanning EMs. This massive range of magnification is really useful, particularly with biological material, enabling selection of areas which have clean surfaces without physical damage, and giving confidence that the areas which are investigated are truly representative of the whole sample. Image formation by the final lens can be distorted by astigmatism if the beam is oval in section rather than circular, and this can be corrected by a set of eight electromagnetic coils (stigmators). The final lens also determines resolution by its formation of a minimum spot size (at best, around 1 nanometre).

Imaging in the scanning EM

If we 'freeze' the scanned probe at any one moment, then we can consider how the beam interacts with the specimen. Unlike transmission EM where the specimen has been suitably thinned to allow the beam to travel through, the bulk nature of specimens in scanning EM prevents this, but the beam does penetrate the specimen surface to a depth of around 1 micrometre, producing a volume of interaction shaped like a teardrop (Figure 18). The collision between the electron beam and the specimen surface involves both elastic and inelastic scattering, and produces secondary electrons, backscattered electrons, X-rays, and cathodoluminescence (visible light). Secondary electrons are formed as a result of inelastic collisions with an atomic nucleus in the specimen where substantial energy loss results from the ejection of loosely bound electrons. Backscattered (or reflected) electrons come from the incident beam via an elastic collision with an atomic nucleus in the sample. It is mainly secondary electrons which contribute the image. Images are formed by modulating the brightness on a cathode ray tube which is scanned in synchrony with the electron beam. The secondary electrons are detected by a detector which is positioned adjacent to the specimen in the chamber, surrounded by a wire grid kept at 200 volts (in order to attract the secondary electrons) in front of a scintillator (kept at 10,000 volts). Having passed through the grid, a secondary electron strikes the surface of the scintillator to produce visible light, which is fed down a light pipe to a photomultiplier which in turn converts the light into an electrical signal. This signal is then used to determine the brightness of the spot on the cathode ray tube to create the image. Thus the more secondary electrons emitted from any one spot on the surface of the specimen, the greater the signal from the detector, and the brighter the area seen at that point in the image on the cathode ray tube. Backscattered electrons also contribute to image formation, and can be imaged alone by either reducing the voltage on the grid in the detector

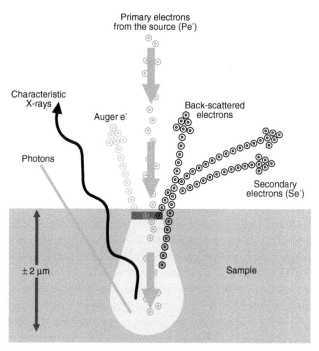

18. The scanned beam forms a teardrop shaped interaction at the surface of the specimen. Secondary electrons (and Auger electrons) emanate from the surface, with backscattered electrons and characteristic X-rays arising from deeper into the teardrop.

(as backscattered electrons have sufficient energy to reach the scintillator regardless), or they can be collected more efficiently by a dedicated solid state detector. Backscattered electrons are useful, as they can be used to map the distribution of elements in the specimen because their energy is directly proportional to the atomic number of the element which they collided with. As they tend to arise from deeper within the surface of the specimen, however, backscatter electrons are not as good as secondary electrons for resolving surface topography. Another signal comes from 'Auger' electrons (which are emitted after interactions between different shells of the atomic structure, named after their discoverer). Auger electrons are only emitted a very short distance from the surface of the specimen, where they are mostly absorbed, but do allow chemical mapping of surfaces to a high resolution (with a dedicated detector). The X-ray emission from a scanned specimen can be used for energy dispersive elemental analysis similar to transmission microscopy. Finally, cathodoluminescence (the production of visible light) can also be detected and is particularly useful in imaging the internal structure of materials such as semiconductors.

Although the introduction of a commercial SEM was much later than transmission instruments, a scanning form of electron microscopy was considered as a possibility by Ruska himself in the very early developments, as methods for making thin specimens for transmission microscopy only emerged later. In 1935 Knoll invented a scanning instrument to study the targets of television camera tubes, to be followed by scanning instruments produced in 1938 and 1942 by von Ardenne and Zworykin respectively. Von Ardenne's instrument was destroyed in an air raid on Berlin in the Second World War. In the USA, the RCA corporation produced transmission microscopes, and were working on scanning instruments, but largely because of difficulties in the vacuum technology of the time, together with the success of surface imaging using replicas in transmission microscopes, never brought a scanning EM to market. Consequently, we have to wait until 1965 for the first commercial instrument, produced by the Cambridge Instrument Company in the UK, as a result of research started in 1948 by Charles Oatley and his team. This was followed just six months later by an instrument produced by the Japanese Electron Optical Corporation (JEOL).

From its earliest days, scanning EM proved to be a source of images that everybody could relate to, regardless of a microscopic or indeed even a scientific background. From early images showing great detail of everyday objects and animals, for example the edge of a scalpel or razor blade or the multiple compound eyes of a spider, the extra information provided by the high magnification was instantly apparent, grasping the attention of the general public in a way that transmission EM images did not (Figure 19). Today, images of bacteria, stem cells, and tumour cells are a regular sight in TV news, documentaries, newspapers, and magazines, usually brightly coloured. False or pseudo-colouring of scanning EM images is useful for highlighting specific features, as well as increasing the overall impact, which can sometimes be a little on the garish side.

In the same way that transmission EM produces signals that generate chemical and elemental information, similar detectors

19. Scanning EM image of a spider: the sort of image which creates public interest in scanning EM, as well as demonstrating the large depth focused information available by this form of imaging.

used were rapidly incorporated into scanning microscopes. Solid state energy dispersive detectors use backscattered electrons to identify individual elements. For best visualization of the distribution of elements specimens are cross-sectioned and polished. This technique was standard for the characterization of powders, particulates, and examination structure in aerospace, automotive, electronics, and semiconductor industries (to name but a few). The more robust nature of scanning EM with respect to specimen preparation and processing has resulted in its incorporation into areas where transmission EM would not have been considered. Scanning EMs are routinely incorporated into assembly lines for quality control, often with automated image analysis. Where a scanning EM specimen was once limited to a few millimetres in diameter, chambers to accommodate the largest silicon wafers (around half a metre) are routine productions. A few years ago a top of the range instrument would be an impressive piece of hardware, with a column weighing in at a quarter of a ton, and a two metre long control desk. Installation would require water cooling, a separate highly smoothed power supply, and sometimes containment in a Faraday cage to exclude any stray electrical field effects. Similar performance is now available in a miniaturized desk top version much the same size as a laser printer which is plugged into the nearest socket available.

Specimen preparation for scanning EM

As a prelude to introducing some more specialized scanning EM configurations, we need to briefly consider the requirements for good imaging of what can best be described as soft stuff—not just biological organisms, but also the use of scanning EM in food science (particularly cream and ice cream), and even the fine structure of ice crystals, which can help in the forecasting of avalanches.

An ideal specimen for scanning EM is a coin—it is dry (no water content), has a hard surface, is made of heavy elements, and is an electrical conductor. These properties ensure that there is no problem with the vacuum in the specimen chamber, a good signal will be produced by the interaction of the electron beam at the surface, and that any electrical charge from the beam can run away to earth due to the conductive nature of the coin. Contrast this with biological tissue or maybe a piece of cheese. They both have around 70 per cent water content, which will evaporate to degrade the vacuum in the specimen chamber at the same time as dehydrating the specimen. The surfaces of our cells (or cheese) will be soft so the electron beam will penetrate more deeply, masking surface detail. The heat dissipated from the beam will also cause damage. On top of this, the light elemental composition of our specimens (made up of carbon, hydrogen, oxygen, and nitrogen) means there will be little useful signal to form images. Finally, as neither cheese nor biological tissue is a good conductor, electrons will build up within the specimen until they have sufficient energy to find a way to earth, resulting in a discharge which generates a burst of signal brightness (known as charging) which obliterates what image there might have been in the first place.

In order to observe biological material (and cheese), we must find a way around these obstacles. Drying a soft surface for SEM is tricky, as the surface tension of the receding liquid layer damages the fragile fine structure. However, liquids have a 'critical point' where both vapour and liquid states coexist, allowing surface tension effects to be avoided if vapour is removed above this critical point. The critical point for water requires fairly extreme conditions, a pressure of around 200 atmospheres and a temperature of 374 degrees C. However, if we follow the standard fixation and dehydration protocols for transmission EM but then transfer the specimen to liquid carbon dioxide, the critical point for CO_2 is 31.1 degrees C, at a pressure of 100 atmospheres. As this point is passed, the meniscus at the surface of the liquid CO_2 disappears. The gaseous carbon dioxide is then allowed to escape, and the specimen is dried free from the distorting effects of surface tension.

We now have a specimen which has retained its fine detail, but the surface is still soft, and its light elemental composition (C,H,O,N) will only produce a poor signal. These limitations are overcome by the deposition of a very thin layer of metal over the surface. For surface replicas made for transmission EM, the metal is evaporated from a point source to produce a shadowing effect, but for scanning EM an even coating is required. This is achieved by a process called sputter coating, in which a target disc of the coating metal is bombarded under vacuum by ionized gas particles (usually argon) which strike the surface and remove atoms which are then deposited on the surface of the specimen. Gold, or an alloy of gold and palladium, produces excellent secondary electron emission, as well as being permanent and non corroding, allowing long storage of samples for future reference. With the highest resolution instruments, however, the actual grains of gold in the coating can be resolved—resulting in the imaging of the coating itself rather than the underlying surface of interest. In these situations chromium or tungsten is used, which is deposited with a smaller grain size (although a reduced secondary electron emission). For studies involving elemental analysis (using backscattered electrons) the specimen surface is made conductive by sputtering a thin layer of carbon.

Labelling for scanning EM

As with transmission EM, small metal particles, such as colloidal gold (conjugated or bound to antibodies), are used as markers to show specific binding sites on biological tissue. In scanning EM the gold label is visualized as a bright sphere rather than a black dot as in transmission EM. Gold labelling and gold coating cancel each other, so carbon coating is usually used so that the colloidal gold stands out. To retain the best surface imaging, chromium coating (atomic number 24) does allow visualization of gold colloid (atomic number 79). Also, as the backscattered electron signal from gold is particularly strong, the backscatter image can be collected simultaneously with the secondary electron signal, so that both can be merged in the computer to give the clearest labelling as well as optimal surface imaging.

Cryoscanning EM

The soft, wet, and largely non-conductive nature of biological material can be made dry and hard by freezing, which is best done as rapidly as possible, converting aqueous water to vitreous ice in milliseconds, as with cryo TEM methods. For cryo SEM the frozen specimen is mounted on a cold stage (cooled by liquid nitrogen) often with a facility to fracture specimens with a cold scalpel blade while still frozen, exposing fresh surfaces for examination. A small sputter coating system allows the freshly fractured surface to be coated with gold in the microscope vacuum system. This approach allows scanning EM observation of material that has been neither chemically fixed nor dehydrated, and is also extremely useful for materials science, food science, and even weather forecasting (see Chapter 8).

Environmental SEM

The preparative requirements for scanning EM would seem to be rather difficult to get round, but the ingenuity of instrument manufacturers should not be underestimated. Let us briefly reconsider the nature of some biological specimens—insects for example, which do have a hard surface (the exoskeleton), and are also conductive when in the living hydrated state. Living ants have been attached to a specimen holder and imaged for a short time until the insect died and became dehydrated, often spoiling the vacuum system at the same time. Botanical specimens, however, particularly those adapted to arid conditions, can be more successfully imaged, again fairly briefly 'au naturel' (i.e. without drying or coating). The use of a low accelerating voltage for the electron beam also helps, reducing penetration of the surface and beam damage, although usually at the cost of resolution.

The simplest solution, however, is to completely remove the requirement for a vacuum environment in the specimen chamber.

Although an electron beam can only be generated in vacuum conditions, once the electrons have been accelerated, they will continue to travel in a directed manner, and can still interact with a specimen even if that specimen is at atmospheric pressure (i.e. no vacuum), or, more surprisingly (and usefully), if the surface is actually wet—even as the surface of a liquid. Liquids are seen as dark opaque shapes because electrons do not penetrate the liquid as light shines through water. The way that electrons can be used for imaging at 'high pressure' (up to 100 per cent humidity) is to generate a vacuum gradient between the electron source and the specimen chamber. Despite the several orders of magnitude difference of vacuum between a field emission source and a wet specimen chamber, computer controlled differential pumping systems together with small pressure limiting apertures between the different levels of vacuum in the column can produce stable imaging in the specimen chamber. Although the electron beam will begin to scatter as soon as it encounters significant gaseous molecules on its journey down the column, and ultimately is lost altogether, there are still sufficient interactions with the wet (and consequently conductive), uncoated surface of the specimen to allow for secondary electron and backscattered electrons to be collected and form images. The usual Everhart Thornley secondary electron detector is unstable in these conditions, and is replaced by a gaseous detection device (GDD) invented by Gerasimos Danilatos in the 1980s, who pioneered the idea of introducing gas into electron microscopes when researching wool fibres in Australia. This unconventional approach made it extremely difficult to fund his research, until a venture capital company in the USA called Electroscan used his designs to build the first commercial 'environmental' scanning EM (ESEM) with Danilatos as director and chief scientific adviser. Once the commercial instrument was produced, its value was soon realized by Philips (then the major European electron microscope producer—now FEI—Field

Emission Industries), which took over Electroscan, and produced a series of ESEMs from the mid 1990s onwards. These instruments can be used in their environmental mode, or pumped out and used 'dry' as conventional instruments. Resolution in environmental microscopes suffers because of the nature of image formation in a 'wet' chamber, but this is more than offset by the ability to put virtually anything that will physically fit into the chamber (including cheese), without any pre-treatment, and get useful information from it—including elemental information from backscattered electrons, as solid state BSE detectors function as normal in a wet chamber. The early uses of environmental SEM in material science included studies of the setting of cement and also (excitingly) to watch paint dry. In much the same way that Oatley deserves credit for pursuing his vision on scanning EM against the prevailing view at the time, so does Danilatos for his persistence in adding an important advance to the family of scanning electron microscopes. Science in general is regularly advanced by dedicated and single minded individuals who have ploughed on with their own beliefs despite a tide of opposing views, and Oatley and Danilatos are shining examples.

Merging the two—scanning and transmission EM

When a thin specimen is viewed in a scanning instrument, the electron beam will pass through it. If the detector is placed underneath (rather than above as usual) an image can be gathered that is similar to that of a transmission EM. From the early years of scanning microscopes, this was well known, but rarely used, as the resolution would be relatively poor. However, as the resolution and quality of imaging in scanning instruments improved, certain advantages to this type of imaging came into consideration. First, the penetrating power of a scanned focused beam is greater than that of the transmission 'flood' beam by a factor of around two to three times (at a given accelerating voltage), allowing thicker sections to be viewed. Secondly, because the scanned beam is only momentarily over any one part of the specimen at any one time, the amount of radiation damage is also considerably reduced. In a transmission EM, the projector lenses which form the image below the specimen can suffer from blurring due to inelastically scattered electrons which cause chromatic aberration. This does not occur in scanning instruments as there are no lenses below the specimen. Also, images can be collected in the scanning EM with digital detectors of almost unlimited pixel numbers (an enormous 32K by 32K in the Zeiss Atlas system). Finally, when the magnification is changed in the scanning EM there is no image rotation as there is in the transmission microscope. All these factors have been utilized in biological samples to produce impressive results in three dimensional reconstructions at the cellular level. Although top of the range scanning instruments are now very close to conventional transmission instrument resolution, sub-atomic detail is still the preserve of specialized TEMs. However, many modern transmission instruments will also incorporate a scanned beam illumination to take advantage of this mode of imaging, termed STEM (Scanning Transmission EM), which allows extended imaging of multiple macromolecules without excessive beam damage, producing large amounts of imaging data to be used for macromolecular reconstructions.

Lifting the lid on the surface

Having collected surface details by scanning EM, the possibility to then strip back this layer and expose underlying structures was a thought that occurred to most microscopists. The electron beam itself lacks the power to do this, but if we replace electrons with something of heavier mass, and provide enough energy, then surface material can be stripped away little by little (termed 'milling'). This is achieved by a technique which uses a beam of focused ions generated from a source in which a tungsten filament coated with gallium is heated under vacuum. The molten gallium flows to the tip of the tungsten needle where gallium atoms are subjected to an electric field which produces ionization and field emission of gallium atoms. Compared to an electron, an atom of gallium is rather like a small cannonball compared to an air rifle pellet. This beam bombards the specimen surface removing atoms and also producing secondary electrons, often sufficient to produce an image on their own. The amount of material removed can be controlled by varying the beam current. At low currents, very little is removed, but the high density of signal allows high resolution imaging. At high beam currents, large amounts can be removed allowing precision milling of the surface micron by micron. This technique is mainly used in semiconductor production and research, using 'dual beam' instruments (FIB-SEMs) which combine SEM imaging with a focused ion beam (FIB) for micro machining. Specimens are imaged with secondary electrons, then 'stripped back' using the ion beam, and imaged again. This can be repeated over and again, often under remote control, until a series of images of successive 'slices' have been recorded. These images can then be reconstructed in the computer to form a three dimensional map. The ion beam itself can be used to produce images, which may lack the ultimate resolution of electrons, but can be extremely useful as they show better elemental contrast of metallic grains in alloys and in mechanical failure and corrosion. As well as material science specimens, blocks of fixed and resin embedded biological tissues and also frozen biological material can be milled to reveal fresh surfaces. Muscle tissue has been reduced by milling to a thin section for transmission EM, with promising results. More recently, an ultramicrotome is inserted into the microscope, and serial sections cut away so that the block face can be successively imaged to produce 3D reconstruction of whole tissues. The introduction of various gases into the specimen chamber also allows chemical modification of the ion beam. More detail is beyond the scope of this brief introduction, except to say that this type of approach forms an important part of production in microelectronics and nanotechnology.

Chapter 7
Magnification by other routes

Scanning probe microscopies

An analogy for scanning probe microscopy might be that of a stylus in the groove of a record. Scanning probe microscopies produce images by moving a finely pointed tip over the surface of the specimen in much the same way as fingertips read Braille. The interaction between tip and surface can provide information not only on the surface topography, but many other characteristics such as magnetic and electrostatic properties, friction, hardness, and even identification of the surface chemistry and its interactions. Resolution of these instruments can be down to atomic level, and the family of instruments which make up the scanning probe microscopies can be considered the third great advance (after light and electron) in the continuous march of microscope technology. Because of their physical interaction with surfaces, these instruments can actually pick up and move around individual atoms, as famously demonstrated by Don Eigler (an IBM fellow) when he spelt out the letters 'I-B-M' with thirty-five xenon atoms on a crystal of nickel in 1989 (Figure 20). This breakthrough is now considered a defining moment in the field of nanotechnology, and the performance and adaptability of scanning probe instruments has proved to be a crucial part in the development of this branch of futuristic science.

20. The letters IBM formed from atoms of xenon manipulated into position on the surface of a chilled crystal of nickel with a scanning tunnelling microscope.

Scanning tunnelling microscopy (STM)

The first commercial scanning tunnelling instrument was produced in 1981, earning its inventors, Heinrich Rohrer and Gerd Binnig, the rapid reward of a share in the 1986 Nobel Prize for Physics (whereas Ernst Ruska had to wait approximately fifty years for his work on electron microscopy to be recognized). Rohrer and Binnig worked at the IBM Research labs in Ruschlikon, Switzerland. One of their novel innovations was to use magnetic levitation to overcome vibration in their system. The analogy of a gramophone stylus and record is not strictly true for scanning tunnelling microscopy because the tip does not actually make physical contact with the specimen. The tip is kept at a very short distance from the surface (in a vacuum), where a voltage difference between the tip and the specimen allows electrons to 'tunnel' between them (as long as the specimen is conductive). Variations in distance between the tip and surface alter the amount of tunnelling current, so that as the tip is scanned over the surface the tunnelling signal is converted into a topographic map (Figure 21). This method of scanning is termed the constant height mode. For scanning, either the tip is moved over the surface of the sample, or the sample is moved while the tip remains fixed. Tip (or specimen) movement is usually driven by a piezoelectric scanning system, which works on the fact that a

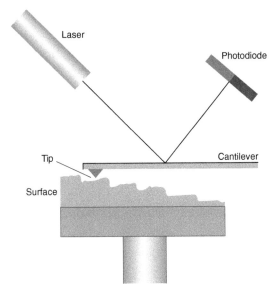

21. As the tip of an atomic force microscope is moved over the specimen surface, the cantilever holding the tip is moved up and down, reflecting the laser light on to a detector to form an image of the surface.

change in voltage across a crystal of lead zirconium titanate (PZT) causes it to alter its physical dimensions. Lateral (x–y) movement is controlled by one system, allowing controlled movement from a nanometre to 100 microns, with a second vertical (z) piezo system capable of height variations from less than one tenth of a nanometre (our old friend the angstrom) to about 10 microns. The change in tunnelling current is recorded as a value on an xy plotter, and this information generates an image produced by computer as a monochrome map of the surface contours, often with 'false' coloration added to emphasize the relative peaks and troughs. Scanning probe microscopy needs time to build up the image as the tip has to physically interact as it travels over the specimen, and it can take several minutes to cover the area of interest.

Tips

The crucial factor for resolution in the scanning tunnelling microscope (and other probe types) is the diameter of the tip. If a tip was scaled up to the size of a football, and passed over a surface of packed golf balls, it would not 'see' the surface contours because the radius of curvature of the football is too great to get in between the surfaces of the individual golf balls—which would need a tip (scaled up) the size of a small bead. The best tips resemble tiny pyramids, which are as rigid as possible consistent with being as sharp as possible. In the early days, making a good tip meant the difference between good or very little resolution, and there are still a variety of methods, depending on the type of scanning probe required, and its particular application. Tips for scanning tunnelling are made from metals such as tungsten or platinum-iridium alloy, by half cutting through wire and then pulling it apart to leave a sharp end, or etching the end with acid. Nanotechnology is used to produce tips, with a silicon wafer being bombarded in a scanning electron microscope with a beam of electrons to induce molecular growth from gas in the chamber to form tips. Alternatively, 'molecular' tips can be made by assembling carbon nanotubes onto a pre-existing tip. Scanning tunnelling tips can also be modified with other chemicals to produce reactions between tips and specimen which allow analysis of the chemistry of the surface under investigation as well as the structural properties. Modified tips can even be made specific for the bases of DNA, enabling guanine, cytosine, thymine, and adenine to be directly identified *in situ*, something never achieved in any other form of microscopy.

Atomic force microscopy

The atomic force microscope was developed from the scanning tunnelling instrument, with a view to overcoming some of its limitations, mainly the reliance on electron tunnelling, which limited its use to conductive samples, and also required a high vacuum environment. Again, Gerd Binnig was involved, this time with Calvin Quate, who had previously been involved in the invention of the scanning acoustic microscope in 1973, and Christoph Gerber from the University of Basel, another pioneer in scanning probe microscopy. The first AFM was announced in 1986, followed by the first commercial instrument in 1989. The atomic force microscope does not rely on electron tunnelling, and consequently can be used for non conducting surfaces, and can operate in both liquid and gaseous environments. The tip makes direct contact with the specimen surface (or very nearly so), measuring the small force between the tip and surface resulting from the nature of the chemical bonds at the surface (hence atomic force). Just how far the tip is moved up and down is usually measured by reflecting a laser beam off the opposite end of the cantilever which bears the tip. The nature of this interaction produces better topographic contrast for more accurate surface mapping, and also resolution is improved compared with scanning tunnelling microscopy. Consequently, AFM has become the main scanning probe instrument for nanotechnology.

The atomic force microscope has different modes of sampling the surface. For frictional force, the tip is dragged across the surface, resulting in torque variations that are measured at the cantilever on which the tip is mounted. A magnetically susceptible probe is used to measure surface magnetic fields, and other tips allow measurement of surface electrostatics and even temperature, with the tip acting as a tiny resistance thermometer. If the tip is likely to deform or become stuck to the surface under investigation, it can be oscillated in the vertical plane in a 'tapping' mode, generally used in air or liquid environments, and particularly for biological materials such as living cells. Atomic force microscopy took some time to accommodate to living materials, as originally a single scan could take several minutes to collect, and the surface molecules of cells would have changed position many times in that period. Scanning electron microscopy is the accepted standard for surface morphology of cells, as the depth of focus available can cope with more extreme topographies than atomic force instruments, but because atomic force can work in aqueous environments, it is possible to use it for living cells. Modern scanning probe instruments will often provide several different modes within the same instrument, including aspects of light microscopy such as near field optical scanning microscopy (NOSM) and micro tools for nanofabrication such as micro-writing devices (nanolithography), indentation probes providing exact positioning and force control, all in a specimen chamber in which both the temperature and gaseous environment can be precisely controlled. This type of scanning probe microscopy has made it possible to investigate a surface phenomenon termed surface plasmon polaritons (SPPs for short), which are surface electromagnetic waves that propagate between the interface of a metal and a dielectric (insulator). More explanation of SPPs would require a VSI on surface physics, but suffice it to say, the scanning plasmon near field microscope has made it possible to work towards practical exploitation in the applications of SPPs (which make it possible to 'package' light in smaller quantities than ever before) in optics, data storage, solar cells, chemical cells, and biosensors.

Near field optical scanning microscopy

Also known as SNOM, scanning near field optical microscopy, the inclusion of 'optical' in the title of this type of microscopy suggests that we should have placed it into previous chapters, but the technology came about through scanning probe precursors. Before we look at the hardware involved, we should briefly consider what is known about a second signal that is undetected in conventional light microscopes which work with light rays that are propagated into the 'far field' (more than a wavelength away from the surface of the specimen) and are subject to Abbe's diffraction limitation with resolution of 200 to 300 nanometres. In the 'near field', however, where the object to lens distance is less than a wavelength, the images which are formed are not subject to diffraction, and resolution can be improved considerably, to around 100 nanometres. The near field signal is also called the evanescent wave, which decays exponentially, making it difficult to image, and limiting it to surface detail only.

The probe used in NOSM usually comes from a laser. It is limited in diameter by an aperture, or by a sharp probe that reflects light onto the specimen. The mechanics of the instrument work in the same way as the atomic force microscope, with the light probe tip scanned over the specimen surface, and a detector to receive and analyse the signals from the specimen (Figure 22). A computer drives the system and displays the surface images on a monitor. The tip of the light probe is kept a constant height from the specimen surface, like the constant height mode of the scanning tunnelling microscope, but with light as the medium of interaction rather than electron tunnelling. A microscope that uses light which is not subject to the diffraction limits

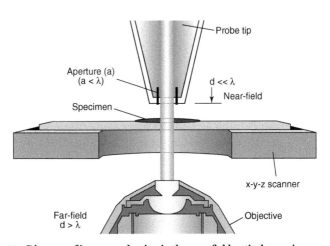

22. Diagram of image production in the near field optical scanning microscope. The near field image is propagated only at sub wavelength distance ($d \ll \lambda$), whereas the far field image is available at distances greater than the wavelength of light ($d > \lambda$).

of conventional microscopy would appear to be a natural progression, but despite the enhanced information it carries, the rapid decay of the evanescent wave signal (within a wavelength of light) presents a hurdle yet to be cleared by current technology, although, as we shall see in the section 'Superlenses', some progress is being made. On a more practical level, near field imaging is expensive and difficult to set up. It is also relatively slow in image collection, and has a best working resolution of around 50 nanometres, some five or six times better than conventional confocal microscopes. However, it is established as a day-to-day instrumentation, not just as an imaging tool, but also for fabrication and manipulation. Applications are continually evolving, such as precision laser machining, nanometre scale optical lithography, and the use of precise amounts of light delivery to activate chemicals called caged compounds, which react to light by altering their properties. An example of this would be a caged chemical label inserted into a cell, which remains inactive until 'switched on' by a burst of light, at which point it would fluoresce, allowing a sequence of events to be followed. Caged probes are not limited to near field microscopy, and are a widespread tool in fluorescence microscopy.

Superlenses

With the advent of nanotechnology, microfabrication has produced novel manmade constructs called metamaterials which exhibit entirely new properties in terms of their effect on light, effects which are not found in conventional materials, or even in nature itself. Early in the 21st century, a chance observation showed that an ultrathin layer of silver on a flat sheet of glass would act like a lens, and from this point, the development of the 'perfect' or 'superlens' began, with the theoretical possibility to image details such as viruses in living cells with a light microscope, bypassing Abbe's diffraction limit. Metamaterials have been produced that make this possible, as they have a property previously unimagined in optics, and not found in nature, which is a negative refractive index. A negative refractive index alters the direction of light passing through in the opposite of what happens in refraction through glass, often termed left 'handed' (Figure 23). In conventional glass lenses, the rays of light which bear the fine details of the specimen but are only present very close to its surface (in the 'near field') cannot be picked up and translated to the far field, where they would be imaged normally. These are 'evanescent' waves, which have been imaged (see section 'Near field optical scanning microscopy') but are limited to surface detail. The goal for metamaterial superlenses is to make the evanescent waves available for normal imaging (i.e. so that they are converted into propagating waves, which are then available for far field imaging). These lenses would work with the visible light range of the electromagnetic spectrum, so that the light microscope (or even the mobile phone camera) would have a resolution close to that of

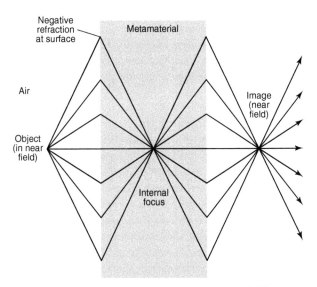

23. **Diagram to illustrate how negative refractive material focuses waves within it, and also at the surface of the lens, to permit imaging at sub wavelength dimensions.**

electron microscopes. Metamaterials with these properties are built by microfabrication, often as a multiple layered 'sandwich' of silicon nitride, gold, and silver layers, all laid down to just a few atoms in thickness. Other metamaterials resemble a stack of irregular miniature venetian blinds in a transparent matrix, with repeated tiny patterns of gold which are smaller than the wavelength of light, and are consequently 'not seen' by the waves moving through the metamaterial that manipulate the passage of light to produce a negative index of refraction. Negative refraction by metamaterials was originally established using microwaves which have a wavelength of several centimetres. To date (early 2014) no one has actually produced a day-to-day working superlens although there are several examples that have produced resolution at around one tenth to one hundredth of the previous light microscope resolution limit (20 to 2 nanometres). These examples usually require laser illumination and image fine details which are actually etched into the front of the lens itself, but they show that it may only be a matter of time before the perfect lens appears. The latest progress has shown that metamaterial optical properties can be produced by a single thin layer of metal coating, indicating the possibility for the propagation of near field information to the far field. This 'perfect' lens was first proposed by Sir John Pendry, a British physicist, in 2000. So far, lenses made of materials of negative refractive indices are still 'near-sighted' in terms of converting evanescent waves to propagating waves, and require the sample to be placed directly on top of the lens, prohibiting imaging of internal cell detail. This leaves us currently with a class of potential superlenses which could actually 'cut through' the diffraction limit to produce single figure nanometre imaging. This is in contrast to the various current working solutions for superresolution that produce unlimited resolution by methods which circumvent diffraction, rather than eliminating diffraction per se. Any method which produces optical images with resolution better than half the wavelength of light falls into a category which is termed optical nanoscopy.

No lenses at all—microscopy with chips

In any one of many videos to be found on You Tube, an enthusiast demonstrates how to dismantle a basic digital camera so the imaging chip (protected by a glass surface) can be exposed. A drop of pond water is then placed directly on the surface, and microscopic organisms in it are visualized as somewhat shadowy images. The key to this 'lens free' imaging is of course the image sensor or chip which would normally receive a focused image from the lens of the camera. Imaging chips work by multiple tiny sensors which convert the light that strikes them into an electrical signal, with each sensor providing the information for one pixel (picture element) in the final image. The smaller and more numerous the sensors are, the greater the detail that can be produced in the final image. Each pixel is the individual component of the image on a computer monitor or flat screen TV. We cannot usually see individual pixels, as they are too small, and although each pixel can only be one colour, they can merge to make thousands of different shades. CCD (semiconductor charge coupled devices) and CMOS (complementary metal oxide semiconductor) imaging chips have revolutionized image recording to the point that microscopy (or photography) recorded on light sensitive emulsion on film has been almost completely superseded. Images collected in this way have a numerical value assigned to each pixel, and are therefore called digital images. Cameras have a digital zoom feature where the image is increased in size by taking central pixels and expanding them. This might appear to be a useful increase in magnification, but the digital zoom has no extra information, just a larger representation of the same detail ('empty magnification').

Digital microscopy

Despite the apparent limitations referred to above, the imaging power of modern chips is such that a sensor with a high number of pixels can pick up massive amounts of information, and this has led to the creation of digital microscopy. A 'standard' digital microscope is really a camera with a CCD chip, mounted on an adjustable stand, with a high resolution zoom lens more akin to a camera zoom lens rather than conventional microscope optics. There is no eyepiece, as the camera displays the specimen image directly to a monitor, and no dedicated light system, as the specimen is viewed with ambient light. The zoom lens allows for a large depth of field, some twenty times that of a conventional optical microscope, so that large irregular specimens can be viewed with all points in focus. The stand enables different angles of viewing, so that accurate two and three dimensional measurements can be made directly, courtesy of the computer control of the whole imaging system. The magnification range can be between one tenth to 5,000, and the whole system is extremely portable. This type of facility is particularly useful in semiconductor production and precision manufacturing, and can be an integral aspect of quality control, positioned above a production line with an automated machine recognition system to check for the consistent appearance of items such as computer chips or pharmaceutical preparations. In a way it could almost be considered as ultra macro photography as much as a microscope, but with a top quoted magnification of 5000 x it is certainly a member of the ever expanding microscopy oeuvre. At the other end of this particular scale, the simple 'USB' microscope plugs into a computer, and a modified webcam and source of illumination can produce satisfyingly magnified images of everyday objects for fifty dollars or less. Simpler (and cheaper still) are the various ways that turn a smartphone into a usable microscope.

Lensless microscopy

Lensless microscopes are instruments produced to specifically use the power of imaging chips directly, rather than to create an image by a lens system as in conventional instruments, which then can be captured digitally in an attached camera. The use of imaging chips, together with computerized image processing and reconstruction programmes, allows fine detail to be obtained by merely placing the object of interest in front of a chip, and illuminating it with almost any part of the electromagnetic spectrum (visible light, lasers, X-rays). A system without lenses is obviously aberration free, and in the case of X-rays (which are notoriously difficult to focus for conventional imaging) makes X-ray microscopy an altogether more accessible technology. Thus, the hardware of lens free microscopes in practice is limited to little more than a viewing platform on which to mount the specimen in front of the imaging sensor. The advantages which accrue from this type of imaging include very large fields of view (as large as the actual sensor itself), which can be 30 square millimetres on a CMOS chip or larger still on a CCD chip. Also, because focus depth is adjusted after image collection, any (or all) levels of focus can be reconstructed, allowing for 3D reconstruction methods. All these apparent benefits, however, do not arise from direct viewing, as the specimen viewed directly from the chip lacks contrast and much in the way of any discernible detail at all. Detail from this type of imaging system comes from the combination of laser or multiple LED light sources and computers powerful enough to perform real time digital processing, as the images begin as holograms produced by the interaction of the illuminating system with the specimen. Although holograms were invented by Dennis Gabor in the 1940s, it was not until the advent of lasers in the 1960s that the first holograms were routinely captured photographically, and reconstructed by shining a source of laser light on the photographic plate, generating a 'magical' three dimensional image. This is because holograms could only be created by illuminating the specimen with coherent light, i.e. light in which all the peaks and troughs of the wavelengths are aligned. Light from lasers is always coherent, whereas light from the sun or a light bulb is not. Very briefly, in holography, the coherent light passing through an object is scattered, then collected and combined with a second unscattered reference beam from the same source. When these two beams come together at the surface of the recording material (originally photographic emulsion, now an imaging chip), they generate an interference pattern. Photographically recorded holograms, when illuminated by an identical beam to the one used to record them, will 'regenerate' a three dimensional image. Whereas photographic holograms are 'frozen' at the instant of recording, the signals received by the imaging chip are reconstructed by suitable algorithms in the computer to produce images of conventional appearance. At the start of the 21st century digital imaging chips also made it possible to generate holograms without the requirement for coherent light, as the hologram is created in the computer. Consequently, large expensive laser sources could be replaced by cheap and compact LEDs.

Images of living cells (which are almost completely transparent) produced by digital holographic microscopy resemble those seen in the phase contrast microscope. More importantly, because the holographic signal is in digital form they can be readily accessed for quantitative measurements, and also for image analysis software. There are other lens free systems such as optical coherence tomography, diffraction phase microscopy, and interferometric microscopy (all of which sadly cannot be considered at any length in this volume), which use similar methodologies to re-form an image from the interaction between illumination and specimen. In some low cost lens free instruments a grid of several LEDs is used for illumination, each set at a slightly different angle to the specimen, to give a different view to add to the information in forming the overall final image.

Lens free on chip microscopy

Using the principles outlined above, lens free, on chip instruments have already developed into lightweight and extremely compact instruments. The set up is a light source (partially coherent and quasi-monochromatic light—most of the radiation confined to a single or extremely narrow waveband). This source illuminates a specimen that is positioned directly on an optoelectronic image sensor array—a system by which light is perceived and translated into electric power. Light scattered by the specimen interferes with the unperturbed background light, creating a hologram that is sampled and digitized by the sensor array (usually a CMOS or CCD chip). As the specimen is positioned on the surface of the chip, the image is produced at unit magnification (i.e. one to one), and each pixel captured by each element on the chip becomes the smallest unit of information (called the pixel function). This determines the resolution of these instruments, and the more pixels that can be fitted into the imaging chip, then the greater the resolution of the final image. In this situation, one might imagine that only the production of improved chips could move things forward, but hologram deconvolution together with a computer image processing technique called 'pixel superresolution' can improve on the limitations of the actual chip dimensions by a significant amount. In 2013, using an light emitting diode source of short wavelength, Aydocan Ozcan and colleagues at the University of Los Angeles reported resolution from a lens free, on chip microscope of better than 200 nanometres—pretty much the limit of conventional light microscope resolution. Ozcan has also developed a lightweight (38 g) and cheap ($10) instrument that images by attachment to a cellphone. The samples are loaded from the side, and illuminated by a simple LED. The incoherent light from the LED is scattered by the specimen to coherently interfere with the background light, creating a lens free hologram on the detector array of the cellphone. Blood cells, platelets, and disease causing micro organisms such as *Plasmodium* (malaria) and *Giardia* can all be suitably imaged for diagnostic purposes. As it is predicted that by 2015 86 per cent of the world's population will own a cellphone, this technology should have a significant impact not only for rapid diagnosis of diseases—a particular problem in third world countries—but also in the prevention of disease by providing a simple and rapid screening of water quality.

Lab on a chip microscopy

As well as the imaging possibilities of on chip devices, there are further possibilities using 'lab on a chip' technologies. Lab on a chip is the term for the miniaturization of any reaction, interaction, or process in chemistry, physics, biology, or material science in which the 'players' are brought together in very small amounts, and interact in extremely limited spaces produced by microfabrication techniques. These are usually hand held (or considerably smaller) devices that produce assays and diagnostic tests using a fraction of materials and resources that were required previously. The production of these microscopic 'test tubes' and reaction vessels allows microfluidics (very small amounts of liquids) and nanofluidics (almost individual molecules) to produce results in liquid chromatography and capillary electrophoresis in chemical applications, and in DNA and RNA sequencing in molecular biology. The ability to control and direct very small amounts of liquid through optically clear materials means that lab on a chip and microscopy on a chip have come together, enabling individual cells to be studied *in situ*, or one at a time, with a high throughput. *Caenorhabditis elegans*, a nematode worm studied by developmental researchers worldwide, has been imaged and measured at a rate of 40 embryos per minute, with a level of detail equivalent to a good x 20 objective lens on a conventional instrument and quite adequate for the morphological information required. It is this sort of automation of specimen handling that makes a level of data gathering possible which was previously unimagined, enabling new levels of experimental design.

Lensless X-ray microscopy

X-ray microscopy has the massive advantage that thick specimens can be easily penetrated relative to other illumination sources. The small wavelength of X- rays is also good for resolution. The main difficulties in conventional X-ray microscopy are the problem of producing lenses that diffract X-ray beams, and the requirement for a suitable source of X-rays. A lensless approach neatly sidesteps the lens problems, but lensless systems still require a coherent source to re-form images from the scattering (diffraction patterns) of the beam interactions with the specimen. Coherent X-rays were only available from large accelerators until it was discovered that if an infrared coherent laser source (wavelength 780 nanometres) is shone through a gas filled tube, atoms in the gas absorb 'bunches' of photons and produce a coherent X-ray beam with a wavelength of 29 nanometres. These X-rays can be generated on the laboratory bench without the need for the access to linear accelerators. The technique for re-forming the images from the scattered X-rays is called ptychography and is similar to that used for other lensless systems, although it was originally developed for improving detail in electron microscope images in the 1960s. There is still the problem of radiation damage suffered by live biological specimens, but this may be circumvented by a recent illumination system that subjects the cell to a pulse of X-rays that lasts for a femtosecond, i.e. one quadrillionth of a second (or 10^{-15}), which not unreasonably claims that the diffracted X-rays are collected before any damage can actually be registered.

Chapter 8
The impact of microscopy

Almost every aspect of our day-to-day existence has been influenced by microscopy. Any attempt to cover everything would fill this entire book, so this chapter is consequently an eclectic selection. Some indication of the universality of microscopy is to view the list of applications in a manufacturer's website. Biomedical applications alone range from bacteriology through pathology to toxicology and neuroscience. In material science, metallurgical microscopes are a major category, and there is an ever increasing requirement for microscopy in both research and quality control of microelectronics and mechanical microsystems. Nanotechnology could be considered an area almost completely dependent on microscopy for its visualization. Amongst the rest of the world of microscopy in general, the following are a few selections, plucked at random amongst a subject which could easily fill this volume and several others.

Microscopy in forensic science

Microscopy in forensic investigation is essentially the search for, and identification of, fragments which may be crucial to a chain of evidence (see *Forensic Science VSI* by Jim Fraser). A crucial element is gunshot residue (GSR), which occurs when a firearm is discharged, as it is deposited on the hands and clothing, consisting of unburnt particles of the explosive charge, along with fragments from the bullet and cartridge case. GSR is deposited up to one and a half metres from the point at which the gun was fired. Particles of GSR vary between 1 micrometre to 20 micrometres, and are identified with a scanning electron microscope which incorporates energy dispersive X-ray spectroscopy for elemental chemical characterization (usually lead, antimony, and barium). This information does not match GSR to a specific gun, however; that depends on ballistic fingerprinting, performed in a special comparison light microscope, in which two bullets can be examined side by side. Other essentials of forensic investigation such as paint analysis are usually carried out by light microscopy (possibly extended to the scanning EM). Sophisticated light stereozoom instruments are used for fibre identification and matching paper fragments. Although there are many other aspects to forensic science, microscopy is at the forefront.

Microscopy in art history—finding the original

Four virtually identical copies of a 16th century painting of 'Christ Driving the Traders from the Temple' are in galleries in Europe. They are in the style of Hieronymus Bosch or Pieter Bruegel, but their relative history and which was the original had never been established. An interdisciplinary research project was started in 2009, using microscopy to characterize the pigments used, how they were ground, and then applied to the wood panels. Cross-sections through an area of a blue garment were compared by polarizing microscopy to show the distribution of azurite and lead white particles. As well as polarizing microscopy, stereomicroscopy was used to analyse the brushwork, along with digital X-radiography and dendrochronology (tree ring analysis of the wood on which they were painted). The analysis revealed that the earliest painting was from 1530, and that the version in Glasgow, signed by Hieronymus Bosch, appears to have been painted eighty years after his death. All seem to reflect the desire for paintings at the time, when it was normal for copies of paintings to be made, to fulfil the demand, in this case, from the Antwerp art market. Microscopy is also a major analytical technique for antique papers and books, glass, ceramics, and stone, as well as minerals and fossils, and throughout art conservation in general.

Microscopy in environmental sciences and mining

Environmental science involves multiple disciplines, from geology, biology, and ecology to toxicology, climatic, and atmospheric studies. Monitoring of the atmosphere, soil, and water systems, as well as changes in flora and fauna, indicates environmental alteration, and also the presence of unacceptable contaminants, such as the identification of asbestos in older buildings. Despite Pliny the Younger noting the poor health of slaves in asbestos mines some 2,000 years ago, asbestos dust was only banned from workplaces in Britain in 1898. Asbestos is established as a cause of respiratory disease and lung cancer, including a rare and intractable condition termed mesothelioma. Asbestos has been in such widespread use that it can turn up in any building, and requires identification before suitable disposal can take place. Light microscopy was used to identify asbestos fibre sizes down to 1 micrometre in diameter, but when electron microscopy was first used, many classes of smaller fibres were discovered. The crystalline structure of the different types of asbestos fibres is characterized using polarized and phase contrast light microscopy, which can be performed on site, to trigger suitable precautions in the removal of suspicious insulation material. Several other asbestos identification methods have been tried, but only microscopy has proved sufficiently reliable.

The source of energy in fossil fuels comes from plant material laid down in coal and oil bearing rocks over geological time scales. Reservoir rocks such as shale and sandstone contain trapped oil and gas between the grains of rock, clays, and other materials. In order to assess the potential yield, rock samples are examined by light microscopy (also possibly extended to scanning EM) for their pore size, and the interconnections between pores. In the exploration of coal, the reflectance of an organic component called vitrinite (which gives coal its shiny appearance) is measured by light microscopy over individual particles to assess the thermal maturity or quality of a particular coal.

Extraction of iron is usually from deposits in which there is a mixture of iron ore (haematite and magnetite) and worthless material known as gangue. Here both reflected and polarized light microscopy is used for classification, and also scanning EM for mineral analysis. Recent advances in backscattered electron sensitivity can separate haematite from magnetite. Reflected light and polarizing light microscopy are also routine tools in the mining of copper, whereas scanning EM with elemental analysis is routine in the extraction of platinum group metals. The search for diamond bearing kimberlite, which comes from deep in the earth, begins with the analysis of hundreds of field samples for garnet, olivine, and crustal xenoliths (rock fragments found enveloped in larger rocks), which indicate geological conditions likely to have produced diamonds. Automated scanning EM with elemental analysis for particular ratios of magnesium and chromium produce data indicating the best statistical probability of areas in a chosen site.

Microscopy in manufacturing

The continuing trend to smaller and smaller components in high precision processes and machines (even before the birth of nanotechnology) has resulted in tolerances and dimensions that need to be accurate to micrometres or even nanometres. Production lines of components need to be constantly monitored for quality control, along with economic efficiency and cost effectiveness. All types of microscopy are standard in virtually all types of manufacturing, particularly in areas such as silicon wafers, transistors, diodes, and integrated circuits. In solar cell production, the deposition of various layers of components for their coverage, and continuity and thickness, are difficult to measure with conventional light microscopy, so optical sectioning provided by confocal microscopy using interference contrast methods is used. Both precision and large area coverage are crucial in the creation of electronic circuits on silicon wafers that can measure in excess of 3 metres, requiring large area coverage by light microscopy and also scanning EM to monitor the efficiency of etching and deposition of circuits.

In large scale manufacturing, microscopy is still an absolute requirement, even if it is not for the inspection of the parts themselves. The requirement for cleanliness in the automotive industry is crucial to reliability, particularly when components are sourced from different suppliers, and cannot be checked once they are installed. In order to check for cleanliness, the finished parts are washed or rinsed, and the liquid filtered to collect any contaminating particles. This process is known as residual dirt analysis, with the last stage being the microscopic examination of the filter itself. This is performed using a light microscope with a specialized scanning stage and digital camera, together with a computer and analytical software which measures the size and number of dirt particles on the filter, which are all displayed automatically on the monitor. Not only are the particles measured down to 5 micrometres, but they are also checked to ascertain if they are reflective (i.e. metallic) or non reflective (plastic), to produce a score for damage potential. The requirement for this type of quality control in manufacturing industries across the board has led many microscope producers to offer off the shelf 'cleanliness analysis' systems.

Microscopy in food and drugs

In food science, the identification of bacteria, both desirable (bionic yogurts) and undesirable (contaminated and/or infectious), is assayed with routine light microscopy and specific staining protocols. Once food is produced, there may be various times of storage, during which polarization microscopy is used to detect the formation of crystals, or the breakdown of the structure of emulsions. The role of microscopy in the discovery and subsequent production of drugs involves a battery of sophisticated microscopical approaches. More than 70 per cent of potential new drugs fail the 'ADME Tox' barrier (in that they are unsatisfactory for absorption, distribution, metabolism, excretion, and toxicology). Light microscopy, particularly fluorescence, where multiple probes can be incorporated, is crucial to show sites of drug binding, drug uptake, and morphological effects. Time lapse imaging can show longer term behavioural responses, and all results are collected using computer controlled image capture for high throughput analysis. Both transmission and scanning EM are routine, particularly for analysis of organelles within the cell such as mitochondria, to visualize membrane alterations which precede cell death.

Once a drug has been found to have suitable activity, an optimal delivery strategy must be found. It will be combined with excipients (a mixture of diluents, binders, solubilizers, and disintegrants). Polarized light microscopy is routine to monitor physiochemical properties, using characteristic refractive indices and birefringence endpoints. Liposomal encapsulation of drugs with membranes of phospholipid molecules which mimic the naturally occurring boundaries of cells protects active compounds against the effects of stomach acid so that they can reach the desired target within the body, and be released with maximum efficacy. Liposomal manufacture requires microscopic sizing, with sizes controlled to between 50 and 500 nanometres. Other advanced drug delivery systems, such as controlled release, are monitored throughout their production by microscopy.

Microscopy and the weather

The advent of low temperature scanning EM led to a study by Bill Wergin and colleagues from NASA in which they collected samples from different types of snow cover found in the prairies, taiga (snow forest), and alpine environments. With snow depths up to a metre, various layers occurred in which the crystals underwent a change in their microscopic shape from the original freshly fallen crystals, to the development of flat faces and sharp edges. It is this metamorphosis of lying snow that determines the likelihood of avalanches, which can be predicted from the crystal structures at various depths. Although scanning EM is hardly available as a routine assay in distant mountain regions, this work helped in the use of microwave radiology investigation of the snow water equivalent in the snow pack, as large snow crystals scatter passive microwave more than small crystals. Smaller and more rounded crystals of snow do not interlock, and can slide more easily over each other, increasing the risk of avalanches.

The origami microscope

More than a billion people in less developed countries suffer from conditions such as malaria (which kills one million a year), sleeping sickness, schistosomiasis, and chagas disease, all of which need to be diagnosed by basic light microscopy, but are rarely identified on site due to lack of expertise and facilities. In 2012, Manu Prakash, an assistant professor from Stanford University, produced an A4 sheet of thin card onto which were incorporated all aspects of a simple but extremely effective microscope (Figure 24), which can be viewed directly, or even act as projection microscope for group viewing. The images produced are of good quality, and adequate to diagnose the presence of the disease causing microbes of giardiasis or leishmaniasis. The flat packed 'foldscope' can be assembled in minutes, has no mechanical moving parts (focus is optimized by gently flexing the whole system), and if used for diagnosis of an infectious disease, it can be incinerated after use, as the total cost is around 50 cents. Samples are mounted on a 3 by 1 inch glass slide (a universal standard) which is slid between the lenses and an illuminating LED incorporated into the card together with its button battery. The foldscope can survive being stepped on, and/or dropped from a five storey building. Perhaps the most novel aspect is the use of spherical glass lenses (at 17 cents each) about the size of a poppy seed, which were originally

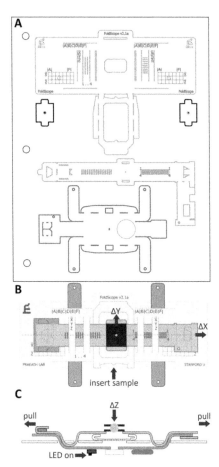

24. A foldscope, shown as manufactured on an A4 sheet (A), then viewed after assembly, from above (B) and the side (C).

mass produced as an abrasive grit. The spherical nature of the lenses, which requires the eye to be placed close, can produce magnifications of below 100 up to 2,000 times, without the need for immersion oil (seriously reminiscent of the instruments of

Antoni van Leeuwenhoek). Even fluorescence microscopy can be incorporated by the use of a coloured LED illuminator and 3 mm square filter inserts, keeping the cost down. The foldscope can also be coupled to a conventional smartphone for image capture, and their transmission worldwide (telemicroscopy), allowing remote live video consultation for expert diagnostic opinion if required. To quote the last sentence of the original paper, 'Our long term vision is to universalise frugal science, using this platform to bring microscopy to the masses.'

At one end of the scale, astronomers search the heavens for new information about the universe, whilst at the other end, microscopists chase atoms and molecules to study defects in crystals or the basic processes of life. These investigations may be separated by more than twentyfold orders of magnitude, but are nevertheless driven by the same insatiable curiosity of the human psyche to explore beyond the vision of our own eyes.

Further reading

Much of the microscopy literature is in textbook form, often in large tomes exceeding 500 pages, but still readable for those who want to explore the subject in greater depth. They are usually devoted to either light or electron microscopy—probably as it would require in excess of 1,000 pages to do justice to both in a single volume.

For light microscopy, there are two recent books, both of which cover all the basics mentioned here in Chapters 1–4.

Douglas B. Murphy and Michael W. Davison, *Fundamentals of Light Microscopy and Electronic Imaging*, 2nd edition (Wiley Blackwell, 2012). Comprehensive and beautifully illustrated—the 50-page glossary covers every term in light microscopy. The title may be slightly misleading, in that 'electronic imaging' is the collection and processing of images from light microscopy alone, and does not cover any electron microscopy.

Ulrich Kubitscheck (ed.), *Fluorescence Microscopy: From Principles to Biological Applications* (Wiley VCH, 2013). High level textbook, 600 pages, basics and great depth of coverage of fluorescence microscopy, although back cover material states that the book 'keeps the non-expert in mind'.

At a more basic level, and with younger readers in mind, but still well worth a look for its range of subject coverage, with every page overflowing with coloured illustrations and also containing useful tips for anybody about to buy and/or start using a microscope at home.

Kirsteen Rogers, *The Usborne Complete Book of the Microscope* (Usborne London, 2012).

Further reading in electron microscopy also tends to come in 500-page plus textbooks, but again they generally cover the basics at a readable level before moving on to more specialized content. A good example of this is the chapter on 'Fundamentals of Scanning Electron Microscopy' by Rob Apkarian, David Joy, and others, and which provides an account by pioneers of the field, to be found in W. Zhou and Z. L. Wang (eds), *Scanning Microscopy for Nanotechnology: Techniques and Applications* (Springer, 2007). The book also contains a comprehensive coverage of all aspects of non-biological SEM.

For TEM, J. Thomas and T. Gemming, *Analytical Transmission Electron Microscopy: An Introduction for Operators* (Springer, 2014), 348 pages and 238 illustrations, may also appear to be a hardcore text, but again the early chapters cover the basics in a readable manner.

Further depth of reading in scanning probe microscopy also requires reference to textbooks, e.g. B. Blushan (ed.), *Scanning Probe Microscopy in Nanoscience and Nanotechnology* (Springer, 2012), a substantial 956-page work with 300 illustrations, but again with early chapters on the basics at a fairly readable level.

Web resources

Most major instrument manufacturers have websites with a wealth of information that is well illustrated and easy to access, often with simulations of the imaging technology (e.g. Olympus Microscopy Resource Centre—<http://www.olympusmicro.com/primer>), allowing for a 'feel' of how changing parameters in the instrument affects the production of the image. Similar sites are at Carl Zeiss Microscopy Online Campus <http://www.zeiss-campus.magnet.fsu.edu> and Leica Science Lab <http://www.leica-microsystems.com>.

Many university and research institute websites will often feature the sorts of microscopy for which an individual department has a particular expertise. Choose a department (e.g. Life Science or Metallurgy), and go to facilities for information on the latest specialist techniques. For example the John Innes

Institute—<http://www.jic.ac.uk/microscopy/links.html>—also has a link to a page with twenty-four further sites covering every aspect of biological microscopy. Failing that, use a major search engine, e.g. input History of Electron Microscopy, and material from esoteric sources appears such as Supplement 16, Advances in Electronics and Electron Physics—'The Beginnings of Electron Microscopy' by Peter Hawkes, one of the founding fathers himself, and made available by the publisher (Academic Press).

Last but not least, and probably simplest of all—'You Tube and Microscopy' will produce a virtually unlimited series of short video material from manufacturers, academics, and enthusiasts on every aspect of the subject.

Index

Index